THE IMPACT OF RATE-OF-RETURN REGULATION ON TECHNOLOGICAL INNOVATION

To Catherine

The Impact of Rate-of-Return Regulation on Technological Innovation

MARK W. FRANK
Sam Houston State University

Routledge
Taylor & Francis Group

LONDON AND NEW YORK

First published 2001 by Ashgate Publishing

Published 2017 by Routledge
2 Park Square, Milton Park, Abingdon, Oxfordshire OX14 4RN
711 Third Avenue, New York, NY 10017, USA

First issued in paperback 2017

Routledge is an imprint of the Taylor & Francis Group, an informa business

British Library Cataloguing in Publication Data
Frank, Mark W.
    The impact of rate-of-return regulation on technological innovation. - (The Bruton Center for Development Studies series)
    1.Rate of return 2.Technological innovations - Economic aspects
    I.Title II.Bruton Center for Development Studies
    338'.064

Library of Congress Control Number: 2001022163

ISBN 13: 978-1-138-26385-7 (pbk)
ISBN 13: 978-0-7546-1609-2 (hbk)

Contents

List of Figures

List of Tables

Acknowledgements

Either by advice, encouragement, or attitude, a great many people have knowingly and unknowingly contributed to this accomplishment. I will cite a few of these people here and ask that those I fail to mention to forgive the omission and take joy in the outcome nonetheless.

I begin by thanking Dr. Barry J. Seldon, Dr. Brian J. L. Berry, Dr. Euel Elliott, and Dr James C. Murdoch; I am proud to have worked with each of them. I am also deeply appreciative to my parents, my sister Tamy, Doris Rush, my parents by marriage Rose and Maurice, and the rest of my family; they have all given me much more encouragement, love, and respect than I deserve.

Finally, I am most grateful to my wife, Denise, who did her best to keep me safe from bad grammar, long sentences, and other maladies. She endured many long nights and weekends alone while I worked, yet always seemed to have time to listen to my problems and concerns. Words alone cannot express how truly thankful and indebted I am to her.

Introduction

The impact of regulation on the performance of firms with market power has long been an important and controversial topic. Although the conventional wisdom among policymakers has generally favored the regulation of such firms, economists have remained skeptical of these efforts. Joseph Schumpeter, for example, is well known for arguing that monopolistic firms are in fact primarily responsible for many of the technological advances and much of the economic progress enjoyed by society. According to Schumpeter:

> Perfect competition is not only impossible but inferior (to monopolistic practices), and has no title to being set up as a model of ideal efficiency. It is hence a mistake to base the theory of government regulation of industry on the principle that big business should be made to work as the respective industry would work in perfect competition (Schumpeter 1950, 106).

As a result, if firms with market power do possess unique and substantial advantages over other firms, then ideally, the regulation of monopolistic firms should avoid to whatever extent possible the extirpation of these advantages. This, however, is not and has not been the case. Richard Posner (1975), for example, argues that in fact the government regulation of monopolies is a larger source of social costs than the costs from the monopoly itself. Similarly, Robert Hahn and John Hird (1991) have found supportive empirical evidence indicating that the efficiency cost from regulation is large, while the benefits from regulation are positive but small.

The general implication from these authors is that the various forms of regulation have costs as well as benefits, and in some cases, the costs of regulation may outweigh their benefits. The contention here is that this may occur precisely because regulation causes inferior technological investment, and in the long run, inferior technological advancement.

1

Hence, by directly impeding technological advancement, regulation may indirectly exacerbate the problem for which it was implemented.

The Research Question

The primary research question motivating this study concerns the impact of government regulation on the innovativeness of firms with market power. More specifically, there are two general issues to be addressed:

1. Will regulated monopolies engage in more or less technological innovation than unregulated monopolies?
2. If unregulated monopolies do engage in more research and development than regulated monopolies, will social welfare be greater with unregulated monopolies (or less regulated monopolies) than with regulated monopolies (or more regulated monopolies)?

The former will consume most of this study's attention, while the latter will be addressed at the end.

For the purposes of this study, "government regulation" is limited to rate-of-return regulation, and "monopolies" are defined as firms with market power, or more specifically, the power to affect the market price. Hence, market price must be a function of both industry output and firm output:

$$P\left(Q\left(\sum_{i=1}^{n} q_i\right)\right),$$

where n is the total number of firms in the industry, P is market price, Q is market quantity of output, and q_i is output from the ith firm.

This of course is a significant twist on the older and more common question concerning the comparative innovativeness of competitive firms versus monopolistic firms. Here again, Schumpeter plays an important role, since the idea that there is a positive relationship between monopoly power and innovation is largely attributed to him.[1] F. M. Scherer in fact notes that Schumpeter "brought into the main stream of economic discourse the question of what market structures were most favorable to technological change and hence economic growth" (Scherer 1992, 1416).[2] Of course, in the fifty years since Schumpeter's time, research into the comparative innovativeness of different market structures has flourished.

Nevertheless, the question here specifically involves the innovativeness of firms with market power, and to what degree regulation impacts their innovativeness. While the literature on regulation is certainly quite vast, it is not connected well with the equally vast literature on technology and technological innovations. For example, although the classic 1962 article by Harvey Averch and Leland Johnson provides an important critique of rate-of-return regulation, none of the implications therein regard technical change or technological investment. In fact, in the rate-of-return literature, only V. Kerry Smith (1974) has attempted to model the implications of rate-of-return regulation on technical change.[3]

Unfortunately, this lone attempt is at best a limited effort. Hence, the primary contribution of this research is to provide a systematic and more comprehensive investigation of the impact of rate-of-return regulation on the technological innovations of firms. This will be accomplished both through economic modeling of firms with market power that are subject to rate-of-return regulation, as well as through empirical testing of the general propositions which follow from this theoretical modeling.

Outline

Chapter 1 provides both a classification of rate-of-return regulation as part of a class of regulation methods, as well as a brief overview of the historical development of regulation in general, and rate-of-return regulation in particular. In addition, section 1.3 of Chapter 1 provides an overview of the process and players of a typical rate hearing.

Chapter 2 looks in detail at the theoretical model of rate-of-return regulation put forth by Harvey Averch and Leland Johnson, as well as some of the key extensions and criticisms of this model. An entire chapter is devoted to this topic not only because of its importance in the regulation literature, but also because it serves as the general foundation for the theoretical models that follow.

Chapter 3 is the first of the two theoretical chapters, and the first section of Chapter 3 (section 3.1) presents an augmented version of the Averch-Johnson model. Although this model is quite similar to many of the models found in the literature, it contains the important addition of research and development as a factor of production. In addition, an extension of this augmented Averch-Johnson model is presented in section 3.2 in which R&D is allowed to augment capital and labor separately. The effect of research spillovers from other firms is also considered in this later model.

Chapter 4 provides an analysis of research joint ventures when rate-of-return regulation is present. The model used here is a symmetric two-stage Nash equilibrium duopoly model of cost reducing R&D, in which firms choose either to cooperate or compete in the research market during the first stage, and cooperate or compete in the product output market in the second stage. The chapter is divided into the three scenarios that are examined: section 4.1 looks at the case where firms compete against each other in both the research and output stages, section 4.2 analyzes the case where firms form a research joint venture in the first stage but remain noncooperative in the final stage, and section 4.3 looks at the case where the firms cooperate in both R&D and output.

In Chapter 5 a translog cost function is estimated using a sample of ten privately owned electric utilities from the state of Texas. This empirical evidence is then used to test the key propositions from the previous chapters regarding the effect of rate-of-return regulation the innovativeness of firms.

The final chapter provides a brief summary of the findings from the previous chapters, an analysis of the social welfare implications from the key theoretical model of rate-of-return regulation, a general concluding statement, and a statement detailing the limitations of the study and the findings as presented.

Notes

[1] To be completely accurate, although Schumpeter uses the term "monopoly" in his discussions, he defines a monopoly by size, not by an ability to influence market price. Hence, as he defined it, "monopoly really means any large-scale business" (Schumpeter 1939, 1044).

[2] For similar attributions, see the important works by Kamien and Schwartz 1982 or Nelson and Winter 1982.

[3] See chapter 2, section 2.3.4 for a detailed discussion and analysis of Smith's model.

1 Rate-of-Return Regulation

The scope of government involvement in, and regulation of, the economy is exceedingly diverse and difficult to characterize succinctly. The regulation of public utilities, however, is considerably easier to describe. Typically, the term "public utility" is defined as a specific set of industries ranging from energy and communications to transportation. Charles Phillips, for example, notes that there are two general connotations: a broad definition which includes electricity, natural gas, telecommunications, water, sewage, airlines, bus and motor freight carriers, gas and oil pipelines, railroads, and water carriers, and a narrow and more traditional definition which includes only electric, telephone, and gas related industries (Phillips 1988, 4).

Using the traditional definition of public utilities, we see that their regulation is commonly based on statutes and ordinances from local, state, and federal governments, and enforced by state and federal regulatory commissions. On the federal level, there are four regulatory commissions with jurisdiction over the interstate activities of public utilities: the Federal Communications Commission (FCC), the Federal Energy Regulatory Commission (FERC), the Nuclear Regulatory Commission (NRC), and the Security and Exchange Commission (SEC).[1]

The FCC was created by the Communications Act of 1934, and is responsible for the regulation of interstate telephone and telegraph service, and radio and television broadcasting. The FERC was established in 1977, and operates as an independent regulatory agency within the Department of Energy. It is responsible for the interstate regulation of electric energy, natural gas, and oil pipeline rates. Many of these duties were inherited by the FERC from the Federal Power Commission (FPC). The NRC was established in 1974, and is responsible for the regulation and licensing of nuclear energy. The SEC was established in 1933 and is responsible for the regulation of new securities, some aspects of stock exchanges, and the holding companies of electric and gas utilities.

State regulatory commissions, while collectively much more powerful with respect to the regulation of most public utilities, are, nevertheless, harder to characterize. In the state of Texas, for example, the Public Utility Commission of Texas is responsible for the regulation of electric utilities, electric cooperatives, and telephone service, while the Texas Railroad Commission is charged with the regulation of natural gas pipelines and municipal gas utilities. See table 1.1 for a brief summary of these federal and state of Texas regulatory commissions.

Most importantly, the type of regulation imposed on most public utilities by these regulatory commissions is rate-of-return regulation. Briefly, with rate-of-return regulation, the commission must first determine the appropriate level of revenue based on a "fair" return on the firm's capital investments and expenditures. Once this revenue requirement has been determined, it must decide on an appropriate pricing structure which will generate that level of revenue. The next section will look briefly at this use of rate-of-return regulation in the electric, gas, and telephone utility industries. This is followed by a concise legal history of the development of regulation and rate-of-return regulation in the United States, and by a section in which the issues, nature, and processes of a typical rate hearing are investigated.

1.1 Rate-of-Return Regulation in the Gas, Telephone, and Electric Industries

Considering just the gas, telephone, and electric utilities, we see a very similar development in the use of rate-of-return regulation: generally, regulation began in the 1930s with a rather stable regulatory environment, until sometime after the late 1960s, when a process of deregulation took place and the existence of rate-of-return regulation was threatened. As Phillips states, "up until about 1968 - the tremendous expansion of the entire public utility sector was accomplished in a favorable and supportive government" (Phillips 1988, 11). Thereafter, the regulatory environment underwent a series of significant changes, particularly in the late 1970s and early 1980s,

Table 1.1 Selected Regulatory Commissions, 1995

Agency	Established	Number of Commissioners	Years in Term	Employees (full-time)	Expenses (FY1995)	Jurisdiction
FCC	1934	5	5	1,827	$210,130,000	Interstate radio, television, telephone, telegraph, and satellite communications.
FERC	1977[a]	5	5	1,463	163,071,000	Interstate electric power, natural gas and oil pipelines.
NRC	1974[b]	5	5	3,418	463,972,000[d]	Nuclear energy and research.
SEC	1934	5	5	2,300	196,290,000[d]	Electric and gas holding companies, securities, and stock exchanges.
Texas PUC	1975	3	6	220	11,815,700	Intrastate electric power, electric cooperatives, and telephone.
Texas RC	1891	3	6	840	5,110,200	Railroads, intrastate gas, natural gas, and oil.
Texas NRCC	1993[c]	3	6	45	n/a	Sewer and Water companies.

Source: National Association of Regulatory Utility Commissioners (1996), and Phillips (1993).
[a] The predecessor to the FERC was the Federal Power Commission (FPC) which was established in 1938.
[b] The predecessor to the NRC was the Atomic Energy Commission (AEC) which was established in 1946.
[c] The Texas Natural Resource Conservation Commission gained regulatory authority from the Texas Water Commission in 1993.
[d] Fiscal year 1991.

typified by the deregulation of the transportation and communication industries.

In the case of the natural gas industry, for example, federal involvement began with the Natural Gas Act of 1938 which established regulatory control of the industry with the Federal Power Commission (FPC). The regulatory method used by the FPC was rate-of-return regulation with price variations being allowed based only on a cost-of-service basis.[2] The movement towards deregulation began with the Natural Gas Act of 1978, which, among other things, established a process of gradual decontrol of natural gas prices, and moved the powers of regulation from the FPC to the FERC, where it remains today. Thereafter, the FERC began actively encouraging the replacement of rate-of-return regulation with various forms of incentive regulation (see Alger and Toman 1990). The final move towards federal deregulation was enacted with the Natural Gas Wellhead Decontrol Act of 1989.

Unlike other public utilities, *federal* regulation of the natural gas industry has historically been much more important than *state* regulation. Essentially, the gas industry consists of three vertically related sectors: one devoted to the production of gas from underground reservoirs, another devoted to the long-distance transmission of gas through high-pressure pipelines, and a third devoted to the local distribution of gas through low-pressure pipelines (see Phillips 1988). In terms of regulation, the FERC is in charge of the regulation of interstate natural gas pipelines, while the regulation of the intrastate pipeline network is left to the states. This is important because approximately two-thirds of total gas consumption is from gas traded in interstate commerce (see Alger and Toman 1990). As a result, federal regulation has generally been much more important than state regulation, largely because the FERC has jurisdiction over a clear majority of the gas that is consumed. In Texas for example, the Texas Railroad Commission has authority over the intrastate pipelines, but as Jerry Ellig and Michael Giberson note, the commission "imposes few barriers to entry and frequently relies on market forces to set prices" (Ellig and Giberson 1993, 80).

Much like the gas pipeline industry, regulation of the telecommunications industry was relatively stable until the 1980s when various deregulatory efforts were enacted. Telecommunication regulation began in 1934 under the FCC, and lasted for 50 years until the break up and restriction of AT&T to the long-distance market in January of 1984.[3] During this time the FCC and the various state public utility commissions used rate-of-return regulation to determine the appropriate rates for telephone utilities. Even for sometime after the break up of AT&T, the FCC continued to use rate-of-return regulation to set its rates.[4] This lasted throughout the mid-1980s until March of 1989, when the FCC began the use of price caps to directly regulate its rates rather than its return.[5] Price caps are a form of incentive regulation, and according to Braeutigam and Panzar (1993), offer an alternative to rate of return regulation that is particularly advantageous when the regulated firm operates in multiple markets, some of which are competitive and some of which are noncompetitive.

In the specific case of AT&T, the FCC argued that this move to price caps was needed because:

> The Commission has concluded in the past that rate of return regulation does not encourage optimal efficiency. Under traditional rate of return regulation, the carrier's allowed profits are computed from its total invested capital, whether or not the carrier is using capital, labor, operational methods, and pricing in the most efficient manner. To maximize profits, the company has an incentive to manipulate its inputs of capital and labor, without regard to efficiency, and to adopt strategies for investment and pricing based on what it expects the regulatory agency might wish, not necessarily what is best serves its customers and society (FCC 1992, 5322).

Moreover, the FCC noted that this move resulted because of an increasing degree of competition in the industry: "in place of the monolithic Bell System, customers may now select their telecommunications equipment and services from hundreds of suppliers offering an ever-expanding menu of choices" (FCC 1992, 5322).

A similar sequence of events also occurred at the state level. After the breakup of AT&T, the local Bell systems were divided into 161 local exchange and transport areas (LATAs), with each of these being assigned to one of the seven regional bell operating companies (RBOCs). Viscusi, Vernon, and Harrington (1997) report that of the thirty-nine states with more than one LATA, twenty-eight of them moved from rate-of-return regulation to price caps within three years after the breakup. However, these price cap regulatory methods vary widely at the state level, and in fact, most states have never completely eliminated rate of return regulation. Instead, most states enforce price caps plus a sliding scale regulatory scheme which allows the state regulators to adjust prices if the firm's rate of return moves outside a certain range.[6]

In the state of Texas, for example, passage of the Public Utility Regulatory Act (PURA) in 1995 by the state legislature initiated a process of opening up the local markets to competition (see Public Utility Commission of Texas 1999b). Accordingly, on a market-by-market and service-by-service basis, local companies are allowed to have their rates capped for four years without the possibility of reduction by the Public Utility Commission of Texas, the expectation being that by the end of this period, the market would be sufficiently competitive that the Commission could then remove the regulation. To date this removal of regulation has not occurred because, in the words of the Public Utility Commission, "competition has been slow to develop." (Public Utility Commission of Texas 1999b, vii). As a result, in 1997 SWBT and GTE-SW still had a greater than 98% market share, and because of the rate freeze, SWBT earned $288 million in excess profit above the traditionally regulated profit, while GTE-SW earned $22 million in excess profit.

Unlike the telecommunications and gas industries, deregulation in the electric power industry is still in the initial stages, leaving electricity as the nation's last major regulated monopoly. Although a national wholesale electricity market has developed, only fourteen states to date have enacted some form of electric restructuring legislation: Arizona, California, Connecticut, Illinois, Maine, Massachusetts, Montana, New Hampshire, Nevada, Oklahoma, Pennsylvania, Rhode Island, Virginia, and most recently, the state of Texas (Public Utility Commission of Texas, 1999a). In many other states, bills aimed at restructuring the electric industry have been introduced.

The key point of debate in most states, including Texas, involves who should bear the burden of past investments whose costs cannot be recovered with competitive pricing (for a discussion of this debate in

Texas, see for example, House Research Organization 1996). Electric utility companies typically argue that they should recover all of these "stranded" costs because these investments were, at the time of enactment, deemed prudent by the regulators. Those opposed to full recovery argue that electric companies should be allowed to recover only a portion of these stranded costs, with the burden being split between taxpayers and shareholders.

Currently in the state of Texas, the Public Utility Commission of Texas has regulatory jurisdiction over the state's electric utilities. Although the market is open to competition for firms that supply electricity to the regulated utilities, sales directly to end-use retail customers are limited exclusively to legally certified utilities (Public Utility Commission of Texas 1999a). As such, utilities are allowed to charge "discounted rates" which are greater than marginal cost, but less than the allocated cost-of-service rates (Public Utility Commission of Texas 1999a, ES-6).

Texas is rather unique because statewide regulation did not begin until 1975 when Texas lawmakers passed the Public Utility Regulatory Act (PURA).[7] Before this period, local governments and municipalities retained regulatory jurisdiction, although most utility companies faced little if any regulatory oversight. In describing the period between the 1930s and 1975, Jack Hopper notes that "many utility services remained unregulated: cities had no power to control rates in rural areas; industrial electric and gas rates within cities became private contractual matters in the absence of municipal regulation" (Hopper 1976, 779).[8] In the words of the original 1975 legislation:

> This Act is enacted to protect the public interest inherent in rates and services of public utilities. The legislation finds that public utilities are by definition monopolies in the areas they serve; that therefore the normal forces of competition which operate to regulate prices in a free enterprise society do not operate; and that therefore utility rates, operations and services are regulated by public agencies with the objective that such regulation shall operate as a substitute for such competition (*Public Utility Regulatory Act* 1975, 2327-2328).

Since its inception the commission has used rate-of-return regulation to regulate these public utilities. However, with the approval of Senate Bill 7 in June of 1999 by Governor George W. Bush, this is set to change dramatically, with retail competition beginning for most investor-owned utilities in January of 2002.

Consequently, across the three most prominent public utility industries, we see a very similar pattern: initially, a stable regulatory environment thrives for decades, until a process of deregulation begins and threatens existence of rate-of-return regulation. Although the electricity industry is just beginning this process of removing rate-of-return regulation, the gas and telephone industries have largely succeeded in its dismantling (see table 1.2). Thus it is true that the role of rate-of-return regulation in the United States is diminishing, although it certainly remains an important form of regulation not only because of its continuing use in some cases, but also because of its historical prominence.

1.2 The Historical Development of Rate Regulation

To better understand the nature of regulation in general, and rate-of-return regulation in particular, it is beneficial to pursue a momentary digression into the legal development of regulation. As such, the primary constitutional basis from which the federal government has the legal authority to regulate business activity is found in Article I, Section 8 of the Constitution. Often known as the "interstate commerce clause," this section declares that Congress shall have the power to "regulate Commerce with foreign Nations, and among the several States."

Limits on this authority are found in the Fifth and Tenth Amendments of the Constitution. Specifically, the Fifth Amendment states that "no person shall be … deprived of life, liberty, or property, without due process of law; nor shall property be taken without for public use, without just compensation." Since the Supreme Court ruled early on that a business enterprise is a "person," the effect of this Amendment has been to limit the scope and nature of the government's regulatory intervention. Moreover, the Tenth Amendment states "the

Table 1.2 Recent Evolution of Rate-of-Return Regulation

Industry	Controlling Agency	Time of Change	New Form of Regulation
FEDERAL			
Gas	FERC	1978-1989	Incentive regulation
Telephone	FCC	1989	Price Caps
Electric	FERC	–	–
STATE OF TEXAS			
Gas	Texas RC	–	–
Telephone	Texas PUC	1995-ongoing	Deregulation as competition develops
Electric	Texas PUC	January 2002	Elimination of entry barriers coupled with rate freeze

power not delegated to the United States by the Constitution, nor prohibited by it to the States, are reserved to the States respectively, or to the people." The effect of this Amendment has been to delegate all other intrastate regulative authority away from the federal government and to the relevant state governments.

The pivotal case in the right of states to regulate business activity came in the 1877 case of *Munn v. Illinois*. The dispute involved two owners of a Chicago grain elevator, Messrs. Munn and Scott, and an Illinois law passed in 1871 which required owners of grain elevators to obtain a state license, to file their rates for grain storage, and to charge no more than a specified rate for grain storage. This law was in fact only one of the several legislative successes accomplished by the National Grange of the Patrons of Husbandry, which was a powerful agrarian organization founded in 1867, and whose members are more commonly known as the Grangers. Munn and Scott essentially argued that they were engaged in private business and should not be subject to state regulation. This law, they argued, violated the Fifth Amendment of the Constitution because it deprived them of their property and due process. The Supreme Court, however, upheld the Illinois law and ruled against Messrs. Munn and Scott.

In the majority opinion of the Court, Chief Justice Morrison Waite stated that "when private property is affected with a public interest, it ceases to be *juris* private only" (*Munn v. Illinois* 1887, 113). Hence, state or federal regulation was appropriate whenever the industry was "affected" with the public interest. To support this, Chief Justice Waite argued that such powers had been used "in England from time immemorial, and in this country from its colonization to regulate ferries, common carriers, hackmen, bakers, millers, wharfingers, innkeepers ...and in so doing to fix a maximum charge to be made for services rendered, accommodations furnished, and articles to be sold" (*Munn v. Illinois* 1887, 146).

Precisely when an industry becomes affected by the public interest, however, is ambiguous. Kaserman and Mayo (1995) note that numerous cases surfaced in the late 1800s and early 1900s to test this standard, but no consensus emerged regarding which industries were sufficiently affected by the public interest to warrant rate regulation. This quandary was indeed noted in Justice Field's minority opinion:

> If this be sound law, if there be not protection, either in the principles upon which our republican government is founded, or in the prohibitions of the Constitution against such invasion of private rights, all property and all business in the state are held at the mercy of a majority of its legislature. The public has no greater interest in the use of buildings for the storage of grain than it has in the use of buildings for the residences of families, nor, indeed anything like so great an interest (*Munn v. Illinois* 1887, 146-147).

Subsequently, in the 1886 case of *Wabash, St. Louis and Pacific Railway Company v. Illinois*, the Supreme Court ruled that states could not regulate rates of shipments in interstate commerce because the Constitution specifically gave the power to regulate interstate commerce to the federal government. Thus, reformers such as the Grangers who wanted to regulated the railroads, had to turn to the federal government. Hence, in 1887, Congress passed the Interstate Commerce Act, which created the Interstate Commerce Commission whose principal mission was to regulate the railroads.[9]   Similarly, several state regulatory commissions also were created during this period primarily to prevent the railroads from overcharging farmers for grain shipments (see Kaserman and Mayo, 1995).

The determination of proper rates for regulated firms remained a significant ambiguity after the *Munn v. Illinois* decision, and remained so until the decision by the Supreme Court for the *Smyth v. Ames* case in 1898.   In this case, the Court established that a regulated firm was entitled to earn a "fair return" on the "fair value" of the property being used. In the Court's words:

> The basis for all calculations as to the reasonableness of rates ...must be the fair value of the property being used ...The company is entitled to ask for a fair return upon the value of that which it employs for the public convenience ...while the public is entitled to demand ...that no more be extracted from it ...than the services are reasonably worth (*Smyth v. Ames* 1898, 546-547).

Hence, the regulated firm must be allowed to earn a "fair" rate-of-return.

To determine what would be an appropriate rate of return, the Court in *Smyth v. Ames* specified several factors to evaluate the assets of the firm: the original cost of construction, the present cost of construction, operating expenses, expenditures for permanent improvements, the amount and market value of stocks and bonds, and the probable earning capacity of the firm.   Nevertheless, these factors only increased the regulator's problems.   The earning capacity and the market value of stocks and bonds, for example, obviously depend on the earnings level of the firm, but this earnings level depends in part on the rate levels determined by the regulatory commission.   As Roger Sherman argues "setting prices would determine the market values of stocks, and so if the market values of stocks was used as a basis to set those same prices the process would be circular" (Sherman 1985, 180). Thus these criteria proved to be only a marginal improvement.

Further specificity came from the Supreme Court's decision in the *Nebbia v. New York* case (*Nebbia v. New York* 1934). The dispute involved a 1933 law passed in New York that established a Milk Control Board to regulate the prices and practices of milk producers and distributors. Mr. Nebbia was a grocer who decided to sell two quarts of milk and a five cent loaf of bread for eighteen cents, and was sued for violating the Control Board which had established the price of milk at nine cents a quart. The Court ruled in favor of the New York law and argued that virtually all industries were targets for regulatory oversight subject only to the basic constitutional requirement for due process. Writing for the majority, Justice Roberts wrote:

> These decisions must rest, finally, upon the basis that the requirement of due process were not met because the laws were found arbitrary in their operation and effect. But there can be no doubt that upon proper occasion and by appropriate measures that state may regulate a business in any of its aspects, including the prices to be charged for the products or commodities it sells.
>
> So far as the requirement of due process is concerned, and in the absence of other constitutional restrictions, a state is free to adopt whatever economic policy may reasonably be deemed to promote public welfare, and to enforce that policy by legislation adapted to its purpose (*Nebbia v. New York* 1934, 537).

Other remaining ambiguities were clarified in the Supreme Court's 1944 decision in the *Federal Power Commission v. Hope Natural Gas Co.* case. Here the Court ruled that the regulated firm was entitled to a "just and reasonable" return, such that the rate is "sufficient to assure confidence in the financial integrity of the enterprise, so as to maintain its credit and to attract capital" (*Federal Power Commission v. Hope Natural Gas Co.* 1944, 603). The general scheme outlined in *Hope* is based on valuing the rate base by either the original-cost standard or the current replacement cost standard.[10] As Roger Sherman (1983) notes, these schema are the standards that every regulatory commission followed thereafter.

The lasting prominence of the methods first outlined in *Hope* does not necessarily mean that they are sound methods. As the *Hope* case and the other cases demonstrate, the Court has been primarily (if not solely) concerned with determining the appropriate profit level that the regulated firm should be allowed. This is important because in so doing, the Court has tended to ignore the impact of these rate regulations on efficiency incentives. Hence, although these procedures are designed to specify an appropriate "fair" profit, they also may force the firm into an inefficient use of inputs. This is a general statement of the Averch-Johnson effect: regulators impose rate of return regulation on the firm and in so doing, the profit maximizing firm is forced to overuse capital inputs, and therefore is no longer cost-minimizing.

1.3 Rate-of-Return Regulation in Action

The key event in the implementation of rate-of-return regulation is the rate hearing. The principal participants in the hearing are the regulators and the regulated firm. Typically, a rate hearing can be called by either one of these two groups, although other groups such as customers, concerned citizens, competitors, and potential competitors, can be quite influential in the process. It is generally expected that the firm will initiate such a hearing whenever its current rate of return (cost of capital) is significantly more than the rate of return mandated by the commission. Likewise, it is expected that the regulatory commission will initiate a hearing whenever the current rate of return is significantly less than the mandated rate of return.

This suggests a perfectly mechanical regulatory process and a purely objective regulatory commission. While such a description may be a useful simplifying assumption for theoretical purposes, it is important to note that in so doing, some inaccuracies are propagated. George Hilton (1972), for example, argues that federal regulators are faced with vague delegations of authority and staffed with members who typically are professional politicians who serve only for a brief period of time, and commissioners who "are in office for a short enough period so that they do not have the continuity required for effective planning" (Hilton 1972, 48). This skepticism of the motives of regulatory commissions has, of course, a long line of support in economics: George Stigler (1971), Richard Posner (1974, 1975), and Sam Peltzman (1976) all have contributed to the furthering of such ideas. Particularly troubling is the tendency for commission members to move to jobs related to the industries they once formally regulated after their tenure is up. Ross Eckert (1981) finds, in a suggestive sample of 174 commission appointees, that while only 37 percent held jobs in the related private-sector industry before appointment to the regulatory commission, but about 51 percent took related private-sector jobs after their commission appointment was finished.

Once a rate hearing has been convened, it follows two distinct phases: (1) the revenue requirement phase, and (2) the rate design phase. The purpose of the first is to accurately identify and estimate the firm's operating expenses, capital investment, and most importantly, the allowed rate of return, such that the appropriate amount of revenue can be decided on by the regulatory commission. The second phase uses this understanding of the appropriate revenue level to issue a pricing structure which, given quantity of sales, will result in that specific revenue amount. The following two subsections look in greater detail at these two phases.

1.3.1 Revenue Requirement Phase

In the revenue requirement phase, the regulating commission attempts to determine the amount of revenue required to give the firm a fair rate of return on its invested capital. That is, the commission attempts to equate the firm's revenue (R) to its expenses:

(1.1) $R = e + sk$,

where e is the firm's operating expenses, k it the firm's investment in plant and equipment (net of depreciation), and s is the allowed rate of return on that investment. This equation can be rearranged to show that the regulatory commission attempts to set the allowable rate of return (s) equal to the ratio of the firm's revenue net of operating costs $(R - e)$ to its rate base (k). Hence, the commission implicitly attempts to satisfy the following equation:

(1.2) $s = \dfrac{R - e}{k}$.

More specifically, the rate base (k) is defined as the value of the firm's capital stock net of depreciation. This can include the value of tangible or intangible property, and the investment in such property. How this is determined varies widely, with the three most common methods being: (1) the original cost standard, (2) the replacement cost standard, and (3) the fair value standard. The original cost standard requires that the rate base be equal to the original dollar expenditure minus depreciation, the replacement cost standard requires that the rate base be equal to the current market value of the most efficient assets, while the fair value standard is a weighted average of the previous two.

Following the ruling by the Supreme Court in the *Federal Power Commission v. Hope Natural Gas Co.* case, the choice of which method to use to was left to the relevant regulatory commission and its judgment as to which obtains the best results. Viscusi, Vernon, and Harrington (1997) note that currently most regulatory commissions use the original cost method, which is advantageous since it leaves little room for debate and ambiguities because determining what the company originally paid for plant and equipment is relatively straightforward. The well-recognized problem with this method occurs in periods of high inflation, where prices from years before may be much lower than current prices. The replacement cost standard is in part a response to this problem since it allows cost to be determined by current market value.

Operating expenses (e) are defined as wages, salaries, materials, supplies, maintenance, research and development, advertising, charitable contributions, and the like (see Phillips 1988, 169; Kaserman and Mayo 1995, 450). Most expenses are uncontroversial and are easily approved by the regulatory commissions. Debates do occasionally arise about more controversial expenses (such as advertising or research and development), however, and in some instances certain expenses are disallowed. For the most part, such intervention into the firm by the regulatory commission is discouraged. From very early on, the Supreme Court ruled that "the Commission is not the financial manager of the corporation and it is not empowered to substitute its judgement for that of the directors of the corporation" (*Southwestern Bell Telephone Co. v. Missouri Public Service Commission* 1923, 276).

The key task of satisfying equation (*1.1*) is of course the determination of the allowed rate of return (s). The general goal is to set a rate-of-return that will allow the regulated firm just to cover its incremental cost of capital. Since capital is obtained from the issuance of stocks and bonds, the firm's cost of capital is made up of dividend payments from stock and interest costs from bonds (see for example Kahn 1988, Kaserman and Mayo 1995).[11] Hence, estimating the cost of capital depends on an estimation of equity capital costs from stocks, and on debt capital costs from bonds. Kaserman and Mayo (1995) argue that this estimation in reality becomes quite problematic because the equity costs from common stocks depend on the certainty of common stock receipts and an estimation of the future growth of earnings and dividends, both of which can be quite controversial. As Thompson (1991) argues, this problem is in part due to the ambiguity of the guidelines established in the Hope case. In its decision, the court simply stated:

> The equity owner should be commensurate with returns on investments in other enterprises having corresponding risks. That return, moreover, should be sufficient to assure confidence in the financial integrity of the enterprise, so as to maintain its credit and to attract capital (*Federal Power Commission v. Hope Natural Gas Co.* 1944, 603).

As a result, the traditional methods used to calculate the equity capital costs are the Discounted Cash Flow method (DCF), the comparable earnings method, and the Capital Asset Pricing Model (CAPM).[12] The DCF method is often referred to as the "dividend yield plus growth approach" and requires estimating the equity cost (s_e) as the growth rate of dividends per share (g_d) plus the current dividend (d) divided by the current price of the asset (P):

$$(1.3) \qquad s_e = \frac{d}{P} + g_d$$

(see Thompson 1991, and Kaserman and Mayo 1995). Although this method is among the most commonly used, Thompson (1991) argues that the estimation of the growth rate of dividends (g_d) can be problematic because the growth is rate is affected by the earnings rate:

> For example, suppose the earned rate of return was low during the time when data was gathered. It would then follow that the growth in dividends from retained earnings would be low. If earnings increased, then the growth rate would also rise (Thompson 1991, 103).

Hence, the primary drawback to the DCF method is that the growth rate estimate is related to earnings and thus can be uncertain.

The comparable earnings method is on the surface, quite a simple method: a set of companies with a comparable degree of risk is selected, and the average return on equity for these companies is then computed. Thompson (1991) notes that the main advantage to this method is that once the comparable set of firms is selected, the necessary calculations are quite easy. Selecting this set, however, is not so simple. In particular, defining which firms are of comparable risk can "vary widely, depending on the experts' judgement of what factors indicate risk and comparability" (Kolbe et. al. 1984, 41).

The Capital Asset Pricing Model (CAPM) bases the estimate of the firm's equity cost of capital on the rate of return from a risk free assets (such as 90-day U.S. Treasury bills). The typical equity cost using CAPM would be defined as:

$$(1.4) \qquad s_e = r_f + \beta\left(r_m - r_f\right),$$

where s_e is the equity cost, r_f is the return on a risk-free asset, r_m is the expected market return, and β is firm specific parameter denoting risk (see Thompson 1991, and Kaserman and Mayo 1995). According to Thompson, the chief advantage of CAPM is its ability to explicitly account for risk, while its chief disadvantage is that market return (r_m) is difficult to determine. This is due in large part to the variation of inflation over time and the inconsistency of market risk over time, both of which mean that simply averaging past market returns often leads to inaccurate assessment of the expected market return. The firm's risk parameter (β) is also problematic to determine in practice because it tends to vary based on the time interval used. Daily betas vary from weekly betas, and weekly betas vary from monthly betas, and so on.

Once operating expenses (e), capital investment (k) and the appropriate cost of capital (s) have been determined, equation (1.1) can be completed and the revenue requirement can be determined.

Throughout this revenue requirement phase the firm has the inherent incentive to either overstate operating expenses and capital investment, or to overspend on expenses and investment, or to do both, however. In so doing, the firm can manipulate equation (*1.1*) to gain greater revenue, or as Alfred Kahn puts it, to gain "supernormal rates of profit" (Kahn 1988, 27). Padding the rate base, inflating capital accounts, and padding expense accounts, are all symptomatic of this perverse incentive.

The burden of scrutiny over the cost claims of the firm of course lies with the regulatory commission. Nevertheless, as Kahn and others point out: "it would be reasonable to expect regulatory commissions to give these costs the major part of their attention. But in fact they have not done so; they have given their attention to the limitation of profits" (Kahn 1988, 29). This certainly is not a new complaint regarding the enforcement and efficacy of rate regulation, but it is a common one.[13] In any case, given this revenue determination procedure and all its inherent imperfections, once it is completed, then the second phase of the rate hearing, the rate design phase, can begin.

1.3.2 Rate Design Phase

In this phase, the regulatory commission attempts to set a price structure that will achieve the previously determined revenue as specified in equation (*1.1*):　　$R = e + sk$.

This can be quite simple if the regulated firm produces only one relevant service to only one homogenous base of consumers, since, given a quantity sold, there is only one unique price that will generate that specific revenue. Hence, equation (*1.1*) can be rewritten as:

$$pq = e + sk$$

where p is price and q is quantity.

If, however, the regulated firm produces two or more services, or one or more service to a heterogeneous base of consumers, then the process becomes a bit more complex since, given the quantities sold, there is now a range of price combinations that will generate the specified revenue. Thus, the relevant equation becomes

$$\sum_{i=1}^{m} p_i q_i = e + sk$$

where *m* is the number of regulated services provided by the firm. Which particular price combination will be chosen by the regulators is usually unclear.   In such a case, the guiding concern involves determining how these prices should vary across customers and across products. As Kaserman and Mayo (1995) argue, the regulator can use this as an opportunity to cross-subsidize various consumer groups, or to respond to political pressure thus the outcome can in part depend on the political and economic clout of these groups.

Charles Phillips (1988) notes that in designing the rates, the legal guideline is that rates must be "just and reasonable" and avoid "undue discrimination." In practice, the general rule is that price differences are permitted if these differences are based on differences in costs, but are not permitted if these differences are based on differences in demand. That is, rates are usually allowed to vary across consumers if the differences are due to differences in the "cost-of-service" to the consumers. If, however, rates are varying because of differences in demand or the "value-of-service," then this is generally seen as "unjust" price discrimination and disallowed by most regulatory commissions.

Of the two phases in the implementation of rate-of-return regulation, it is this second phase which attracts most of the attention. Perhaps this is a hint of the criticisms that follow. Indeed, the Averch-Johnson critique of rate-of-return regulation, which is presented in the next chapter, deals with problems that arise entirely from the revenue requirement phase. In the years after Averch and Johnson formulated their model, the literature on rate-of-return regulation has dealt almost exclusively with efficiency problems emanating from the revenue requirement phase.[14]

Notes

[1] Regulation by regulatory commissions, of course, was not always the common method. Before the widespread use of commissions, the regulation of public utilities was typically done by judicial decisions and common law, by legislative acts from state legislatures, and by the issuance of franchises by local governments (see for example Phillips 1988).

[2] Cost-of-service is a method used for rate designs after the determination of the revenue requirement has been completed, in which the rates are allowed to vary based on differences in the cost of providing that service to the respective customer (see Phillips 1988, 411-412).

[3] A weak form of rate regulation in the telephone and telegraph industries did begin in 1910 under the jurisdiction of the Interstate Commerce Commission, but was transferred to the newly created FCC in the Communications Act of 1934.

[4] The FCC in fact set AT&T's return at 12.2 percent during these years (see Viscusi, Vernon, and Harrington 1997, 494).

[5] The price cap used by the FCC on AT&T is set so that AT&T is allowed to raise its prices no more than the rate of inflation minus some amount of expected productivity. Of course, AT&T is completely free to lower its prices at any time (see Viscusi, Vernon, and Harrington 1997, 368).

[6] Braeutigam and Panzar (1993) discuss this exact point in relation to state regulation of the local exchange networks.

[7] The act became effective September 1, 1975.

[8] Beginning in 1937, appeals for electric, telephone, and water rates for cities of all sizes were made directly to the courts (see Hopper 1976). This piecemeal approach lasted until the PURA of 1975.

[9] This is not meant to imply that reformers such as the Grangers were the sole forces behind the passage of the Interstate Commerce Act. In fact, Gilligan, Marshall, and Weingast (1989) argue that the railroad companies themselves played a key role in its passage because all other efforts to maintain higher prices and mitigate the negative effects from their aggressive price wars continuously failed. Although Gilligan et. al. explicitly reject the conclusion that the ICA provided the railroads a cartel manager, they do find that while organized antirailroad groups gained important restrictions on railroads particularly in the shorthaul markets, the railroad companies "benefited by earning supracompetitive profits in others" (Gilligan, Marshall, and Weingast 1989, 36).

[10] Section 1.3.1 will cover these standards in greater detail.

[11] Perhaps a better definition is the one given by Alfred Kahn: "the cost of capital...is a weighted average of the separate costs of obtaining funds by sale of bonds, preferred stock, and common stock" (Khan 1988, 50).

[12] Other more recent but less used methods include the arbitrage pricing theory (APT), and the contingent claims method (see Thompson 1991).

[13] On the other hand, it is perhaps an impossible task that regulatory commissions face. Kahn notes that to a large extent, effective regulation of operating expenses and capital outlays would "require a detailed, day-by-day, transaction-by-transaction, and decision-by-decision review of every aspect of the company's operation" (Kahn 1988, 30). This being beyond what is acceptable, leaves regulation as an impossible task.

[14] The exceptions include Davis (1973) and Klevorick (1973), but even these authors expend most of their energy on problems resulting from the revenue requirement phase.

2 The Averch-Johnson Model

The current debate concerning the effects of rate-of-return regulation was initiated in the classic 1962 article by Harvey Averch and Leland Johnson entitled "Behavior of the Firm Under Regulatory Constraint." In this article, the authors claimed that rate-of-return regulation causes the firm to use an inefficient mix of capital and labor. In their own words:

> If the rate of return allowed by the regulatory agency is greater than the cost of capital but is less than the rate of return that would be enjoyed by the firm were it free to maximize profit without regulatory constraint, then the firm will substitute capital for the other factor of production and operate at an output where cost is not minimized (Averch and Johnson 1962, 1053).

This effect of overusing capital relative to the other factor input (labor), has since come to be known as the "Averch-Johnson effect."

Averch and Johnson were not alone in making this argument. Stanislaw Wellisz (1963) and Fred Westfield (1965) also investigated the impact of rate-of-return regulation and came to very similar conclusions. Wellisz was interested in the use of rate-of-return regulation by the Federal Power Commission on natural gas pipeline companies. Although his article is rich in detail surrounding the peak-load and off-peak periods of pipeline companies, the general conclusion of Wellisz is essentially the same as Averch and Johnson: if the fair return on capital is "not much higher" than the cost of capital, then an overexpansion and inefficient allocation of resources occurs (Wellisz 1963, 36).[1]

Westfield argued that firms subject to rate of return regulation will tend to pay inflated prices for capital inputs and as a result, will employ an inefficient capital-labor mix. Hence, "a public utility may not only fail to suffer a decline but actually experience an increase in profit if suppliers of capital goods collude to raise their prices" (Westfield 1965,

441). Others have noted that Westfield's result is a necessary consequence of the Averch-Johnson result; the primary difference between the two articles lies in Westfield's relaxation of many of the more restrictive assumptions of Averch and Johnson (see for example, David Dayan 1975).

Since the Averch-Johnson model is the key model in the literature, the following section will define what is meant by the Averch-Johnson effect, and how Averch and Johnson arrived at this conclusion. Section 2.2 will provide a geometric treatment of the Averch-Johnson model, and section 2.3 will review many of the extensions and elaborations that followed the Averch-Johnson model.

2.1 The Averch-Johnson Effect

Averch and Johnson began their analysis with the simple assumption that production is a function of two inputs, quantity of labor (*l*) and of physical plant and equipment (*k*). Hence, production by a monopolist producing a single homogenous product is defined as:

(*2.1*) $\quad q = q(k,l),$

with

$$\frac{\partial q}{\partial k} > 0,$$

and

$$\frac{\partial q}{\partial l} > 0,$$

where *q* is production output, and positive, nonzero amounts of both capital and labor are needed for nonzero production.

Since the case of a monopoly is considered, the inverse demand function for the firm is defined as:

$$P = P(q),$$

where *P* is price.[2] Cost of production for the firm is defined as

(*2.2*) $\quad c(k,l) = rk + wl,$

where *r* is cost of capital and *w* is the wage rate for labor. It is assumed that the firm has no control over of the price of either of these inputs. Thus profit must be:

(2.3) $\Pi = P(q(k,l))q(k,l) - rk - wl$.

In defining the regulatory constraint, Averch and Johnson choose to ignore both the depreciation of the plant and equipment for this time period, and the cumulative value of depreciation of that plant and equipment.[3] Consequently, the rate-of-return regulatory constraint is defined as:

(2.4) $s \geq \dfrac{P(q(k,l))q(k,l) - wl}{ak}$

where *a* is the acquisition cost of plant and equipment (recall that *r* is strictly the cost of capital), and *s* is the rate of return allowed by the regulatory agency. Accordingly, this allowable rate of return is set to compensate the firm for the cost of holding the plant and equipment. If the acquisition cost is assumed to equal one, then the constraint can be rewritten as:

(2.5) $s \geq \dfrac{P(q(k,l))q(k,l) - wl}{k}$,

or

(2.6) $P(q(k,l))q(k,l) - sk - wl \leq 0$.

The problem for the monopolist is to maximize (2.3) subject to the regulatory constraint (2.6). The Lagrangian expression is thus:

(2.7) $L(k,l,\lambda) = P(q(k,l))q(k,l) - rk - wl$
$\qquad\qquad - \lambda\big[P(q(k,l))q(k,l) - sk - wl\big]$.

The Kuhn-Tucker necessary conditions are:

(2.8) $r \geq (1-\lambda)\left(P + q\dfrac{dP}{dq}\right)\dfrac{\partial q}{\partial k} + \lambda s$,

(2.9) $r > (1-\lambda)\left(P + q\dfrac{dp}{dq_i}\right)\dfrac{\partial q}{\partial k} + \lambda s$,

(2.10) $(1-\lambda)w \geq (1-\lambda)\left(P + q\dfrac{dP}{dq}\right)\dfrac{\partial q}{\partial l}$,

(2.11) $(1-\lambda)w > (1-\lambda)\left(P + q\dfrac{dP}{dq}\right)\dfrac{\partial q}{\partial l}$,

(2.12) $0 \geq P(q(k,l))q(k,l) - sk - wl$,

and

(2.13) $0 > P(q(k,l))q(k,l) - sk - wl$.

It is assumed that the cost of capital must be less than the allowed rate of return $(s > r)$, and that λ varies continuously such that $0 \leq \lambda \leq 1$. If for example $\lambda = 0$, then the rate of return regulation imposed on the firm is either completely ineffective or nonexistent. If, on the other hand, $\lambda = 1$, then the regulation must be completely effective and the allowable rate of return must equal the cost of capital $(s = r)$. As such, the effectiveness of the regulation is indicated by the value of λ.

Rearranging equation (2.8) reveals:

$$\frac{\partial k}{\partial q}(r - s\lambda) = (1-\lambda)\left(P + q\frac{dP}{dq}\right),$$

$$\frac{\partial k}{\partial q} = \frac{(1-\lambda)\left(P + q\dfrac{dP}{dq}\right)}{r - s\lambda},$$

and thus

(2.14) $\partial k = \dfrac{(1-\lambda)\left(P + q\dfrac{dP}{dq}\right)\partial q}{r - s\lambda}$.

Rearranging equation (2.10) similarly reveals:

$$\frac{\partial l}{\partial q}[w(1-\lambda)] = (1-\lambda)\left(P + q\frac{dP}{dq}\right),$$

and

(2.15) $\partial l = \dfrac{(1-\lambda)\left(P + q\dfrac{dP}{dq}\right)\partial q}{w(1-\lambda)}$.

Combining equations (2.14) and (2.15) reveals the marginal rate of substitution:

$$
(2.16) \quad \frac{\partial l}{\partial k} = \frac{(1-\lambda)\left(P + q\dfrac{dP}{dq}\right)\partial q}{w(1-\lambda)} \cdot \frac{r - s\lambda}{(1-\lambda)\left(P + q\dfrac{dP}{dq}\right)\partial q}
$$

$$
= \frac{r - s\lambda}{w(1-\lambda)}
$$

$$
= \frac{1}{w}\left[r - \frac{\lambda}{1-\lambda}(s-r)\right]
$$

$$
= \frac{r}{w} - \frac{\lambda}{(1-\lambda)} \cdot \frac{(s-r)}{w}.
$$

Hence, if regulation is completely ineffective ($\lambda = 0$), then equation (2.16) can be reduced to:

$$
(2.17) \quad \frac{\partial l}{\partial k} = \frac{r}{w}
$$

If, on the other hand, regulation is effective, then

$$
(2.18) \quad \frac{\partial l}{\partial k} < \frac{r}{w}
$$

because $\lambda > 0$ and $s > r$.

Figure 2.1 illustrates the cases where regulation is ineffective ($\lambda = 0$) and where regulation is effective ($\lambda > 0$). When regulation is ineffective, the firm operates at point E, where the marginal rate of substitution of capital for labor is equal to the ratio of the two factor input costs, as is stated in equation (2.17). If, on the other hand, regulation is to some degree effective, the firm is forced to operate at point R where marginal rate of substitution of capital for labor will be less than the ratio of the factor input costs, as is stated in equation (2.18).

Although both points E and R represent the same quantity of output, point R achieves this output only at a higher cost than point E. Thus, the regulated firm's costs are higher than the unregulated firm's costs to produce an equivalent output. As λ increases, it must also be the case that this cost inefficiency also increases. Although Averch and Johnson do not state or explore this point, it is clearly implied, and follows because:

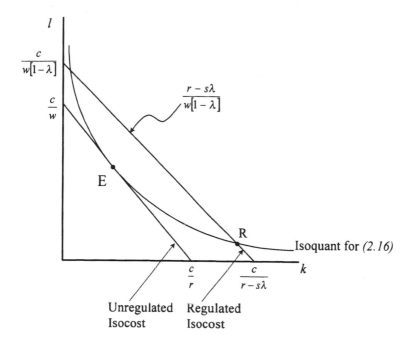

Figure 2.1 The Averch-Johnson Effect

$$\underset{\lambda\to 1}{Lim}\frac{\lambda}{(1-\lambda)}\frac{(s-r)}{w}=\infty .$$

Since equation *(2.16)* is

$$\frac{\partial l}{\partial k}=\frac{r}{w}-\frac{\lambda}{(1-\lambda)}\cdot\frac{(s-r)}{w},$$

then,

$$(2.19)\quad \underset{\lambda\to 1}{Lim}\left[\frac{r}{w}-\frac{\lambda}{(1-\lambda)}\cdot\frac{(s-r)}{w}\right]=\frac{r}{w}-\infty .$$

These cost inefficiencies are solely a result of the overuse of capital relative to the other factor input, labor. This is the Averch-Johnson effect. As they state, the imposition of rate-of-return regulation on the

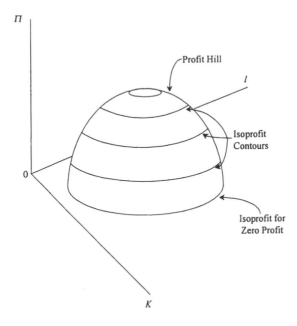

Figure 2.2 The Profit Hill

firm causes a "misallocation of economic resources" in which the firm "has an incentive to substitute between the factors in an uneconomic fashion" (Averch and Johnson 1962, 1068).

2.2 A Geometric Treatment

The original model by Averch and Johnson is primarily mathematical in both setup and execution, relying on nonlinear programming and the Kuhn-Tucker theorem. E. E. Zajac (1970) presents an alternative geometric interpretation of the Averch-Johnson model, arguing that the mathematical methods of Averch and Johnson cause their analysis to be "foreign and uncomfortable" to many of those concerned   with regulation (Zajac 1970, 117). As a consequence, in many discussions of

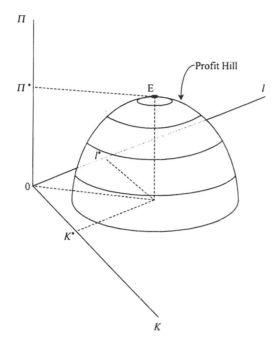

Figure 2.3 Maximum Profit for the Unregulated Firm

the Averch-Johnson model, it is Zajac's geometric treatment that is presented (see for example, Bailey and Coleman 1971; Train 1991; and to a lesser extent Baumol and Klevorick 1970).

The foundation of the geometric treatment is the profit hill, which is presented in its basic form in figure 2.2. The shape of the hill reflects the premise that once output is set, profit is a function of capital inputs and labor inputs. The zero profit contour represents all possible combination of k and l which result in zero profit for the firm. As better combinations of these inputs are utilized by the firm, the firm moves to higher isoprofit contours. It is assumed that there is a single peak to the profit hill and that the surface continually falls away from this single peak, as it is pictured in figure 2.2.[4]

Maximum profit for the firm occurs at the optimal input combination of capital (k^*) and labor (l^*), which is point E in figure 2.3.

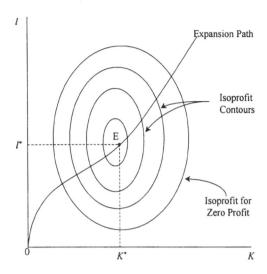

Figure 2.4 Maximum Profit

Figure 2.4 presents this three-dimensional profit hill and point of maximum profit in two dimensions.

Figures 2.2, 2.3, and 2.4 only represent the case of an unregulated firm. To represent the impact of rate of return regulation, a constraint plane is placed on the profit hill. This plane is visualized as a door hinged on the *l*-axis which swings upward along the *k*-axis (see figure 2.5). The key here is that the slope of this constraint plane is determined by the difference between the allowed rate of return and the cost of capital $(s - r)$. If for example, the rate of return exactly equals the cost of capital, then the constraint plane is flat and connects the entire portion of both the *l*-axis and the *k*-axis. As the allowed rate of return (s) increases, the slope of this constraint plane therefore increases and the firm is able to reach a higher isoprofit contour.

Mathematically, this follows because we are considering the case of a monopolist producing a single good (q), with quantities of two inputs (k and l), where profit for the unregulated firm is defined as:

(2.20) $\Pi = P(q(k,l))q(k,l) - rk - wl$.

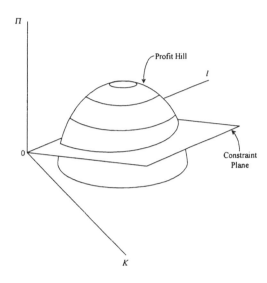

Figure 2.5 The Profit Hill and the Constraint Plane

Likewise, for the regulated firm, the fair rate of return is defined as:

$$(2.21) \quad s \geq \frac{P(q(k,l))q(k,l) - wl}{k}.$$

These are, of course, identical to equations (2.3) and (2.5) used in the Averch-Johnson model. Further rearranging of (2.21) reveals:

$$(2.22) \quad sk \geq P(q(k,l))q(k,l) - wl.$$

Thus, by substituting (2.22) back into (2.20), profit can be restated as:

$$(2.23) \quad \Pi \leq (s-r)k.$$

Since it is assumed that the allowed rate of return is greater than the cost of capital ($s > r$), profits are determined by the quantity of capital and the difference between s and r. Consequently, in figure 2.5 the highest possible profit corresponds to the largest amount of capital on the constraint curve.[5] Hence, in figure 2.6 the unregulated firm produces at point E, while the rate regulated firm produces at point R. For the unregulated firm, profits (Π^*) are maximized by using the optimal combination of capital (k^*) and labor (l^*). The regulated firm by contrast, is restricted by the constraint plane and must increase its use of

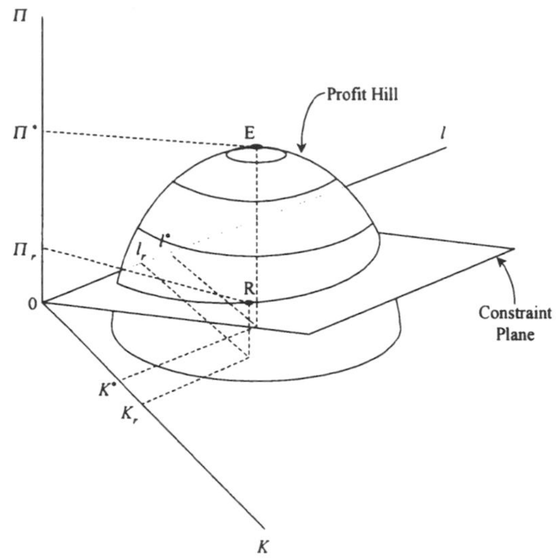

Figure 2.6 Profit for the Regulated Firm

capital to K_r and decrease its use of labor to L_r, which results in a lower total profit (Π_r).

Figure 2.7 presents this comparison in two dimensions and adds the corresponding isoquant and isocosts. Much like figure 2.1, point E represents the operation of the unregulated firm, while point R represents the regulated firm. The constraint curve corresponds to the intersection of the constraint plane on the profit hill from the previous figure. As a result of rate of return regulation, to maintain a given quantity of output, the regulated firm is forced to point R which is on the constraint curve and the same isoquant as point E, but requires a more costly isocost. The Averch-Johnson effect, again.

2.3 Extensions of the Averch-Johnson Model

Several elaborations of the original Averch and Johnson model followed its publication in 1962. Although none of these are particularly

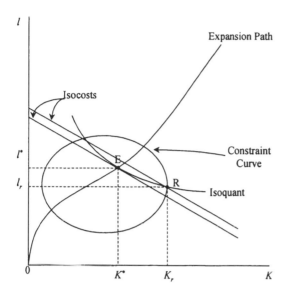

Figure 2.7 The Effects of Regulation

devastating to the primary conclusions of the model, they are important revisions and extensions.[6] These extensions can be grouped into four categories: those allowing for lags in the enforcement of the rate regulation, those dealing with non-profit maximizing firms, those dealing with the case where the allowed rate of return exactly equals the cost of capital ($s = r$), and finally, those that deal with the impact of rate of return regulation on the allocation of technology. The following four subsections deal with these extensions.

2.3.1 Regulation Lags

If rate of return regulation is imposed on the firm, and if it is continuous, a firm operating at any point on the constraint curve other than point R in figure 2.8 could immediately improve its profit by moving to point R. This is so because point R represents the highest point on the profit hill that is intersected by the constraint plane. However, what if a significant amount of time elapses between the actions of the firm and

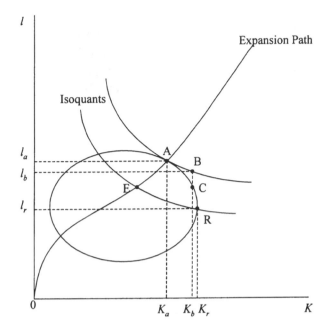

Figure 2.8 Lag Effects

the response of the regulator? Bailey and Coleman (1971) argue that in the presence of a regulatory lag, the firm will still overcapitalize as before, but by some smaller amount, depending on the length of the lag. Hence, the Averch-Johnson effect still persists, but is mitigated to some degree by discrete enforcement.[7]

To show this, Bailey and Coleman, presume that the firm is currently producing at a point such as point A in figure 2.8. At point A, the firm is complying with the regulatory constraint (it is on the constraint curve), but is not maximizing profits. Although point A is cost-minimizing, for the regulated firm, point R would generate greater profits. Thus, if the regulated firm is currently operating at point A, it might consider increasing its use of capital, such that it operates at point B, which is on the same isoquant as A, but is outside the constraint curve and non-cost minimizing.

As Bailey and Coleman argue, the next step depends on the speed of regulatory enforcement. If there is continuous enforcement, the regulator would instantaneously reset price so that the firm operates at a point such as point C, which is on the constraint curve. In this case, point C is certainly better than point A from the firm's perspective because it holds greater profits, although it does not minimize costs. (Of course, increasing capital to point R would be an even greater improvement.) If, on the other hand, enforcement is discrete, then the relevant question is whether the firm should move at all from point A, which is on the regulatory constraint and is cost minimizing. That is, if the time between regulatory enforcement is long enough, then it may be the case that profits are greater when the firm does not overcapitalize (stays at point A instead of moving to point B or point R).

To analyze this, Bailey and Coleman note that according to equation (2.23), profits for the regulated firm can be simply stated as:

$$\Pi \le (s-r)k .$$

Thus profits for the firm operating at point A in figure 2.8, can be defined as:

$$\Pi_a \le (s-r)k_a$$

If the firm decides to remain at this point, then there is no need for regulatory adjustment, and the present value of the firms future profits can be stated simply as:

$$(2.24) \quad \Pi_0 = \Pi_a \sum_{j=0}^{\infty} d_j = \frac{\Pi_a}{1-d} ,$$

where d is the discount rate.

If, on the other hand, the firm chooses to increase its capital from point A to point B, then the regulator responds by forcing the firm to point C, and profits are increased to:

$$\Pi_c = \Pi_a + (s-r)(k_b - k_a).$$

Hence, if regulation occurs at time T, then the move from A to B results first in a profit loss, until the regulator resets price (T) so that the firm is moved to point C, which results in a profit gain to the firm. As a result, discounted profits before regulation for the firm are:

$$\Pi_{ab} = [\Pi_a - r(k_b - k_a) - w(l_b - l_a)] \cdot \sum_{j=0}^{T-1} d_j ,$$

where d is again the value of next year's dollar. Note that $r(k_b - k_a)$ must be greater than $w(l_b - l_a)$ because costs are being increased with the move from A to B (all else equal, B is on a higher isocost than A). Likewise, after regulation, discounted profits are:

$$\Pi_{bc} = \left[\Pi_a + (s - r)(k_b - k_a)\right]\sum_{j=T}^{\infty} d_j .$$

Consequently, the profit steam for the firm moving from A to B, and eventually to point C after regulation is enforced at time T, can be stated as:

$$\Pi_{ac} = \left[\Pi_a - r(k_b - k_a) - w(l_b - l_a)\right]\sum_{j=0}^{T-1} d_j$$

$$+ \left[\Pi_a + (s - r)(k_b - k_a)\right]\sum_{j=T}^{\infty} d_j .$$

Simplifying reveals,

$$\Pi_{ac} = \left[\Pi_a - r(k_b - k_a) - w(l_b - l_a)\right]\frac{1 - d^T}{1 - d}$$

$$+ \left[\Pi_a + (s - r)(k_b - k_a)\right]\frac{d^T}{1 - d},$$

or

$$(2.25) \quad \Pi_{ac} = \frac{\Pi_a}{1 - d} - \frac{\left[r(k_b - k_a) + w(l_b - l_a)\right]\left(1 - d^T\right)}{1 - d}$$

$$+ \frac{(s - r)(k_b - k_a)d^T}{1 - d}.$$

Consequently, overcapitalizing and moving to point B from A, will be worthwhile if and only if profits defined by equation (2.25) are greater than equation (2.24). That is, if $\Pi_{ac} > \Pi_o$, then overcapitalizing is worthwhile. This, of course, will occur if:

$$(2.26) \quad \frac{(s - r)(k_b - k_a)d^T}{1 - d} - \frac{\left[r(k_b - k_a) + w(l_b - l_a)\right]\left(1 - d^T\right)}{1 - d} > 0,$$

where the first term states the profit gain to the firm after regulation is imposed at time T, and the second term represents the profit loss to the firm after its move away from point A, but before the imposition of regulation.

To find the profit maximizing capital increase, equation (*2.26*) is differentiated with respect to *k*. Hence,

(*2.27*) $$\frac{d(\Pi_1 - \Pi_o)}{d(k_b - k_a)} = (s - r)\frac{d^T}{1-d} - \left[r - w\frac{d(l_b - l_a)}{d(k_b - k_a)}\right]\frac{1-d^T}{1-d},$$

which is similar to equation (*11*) in Bailey and Coleman. Consequently, overcapitalization (the Averch-Johnson effect) is worthwhile if and only if:

(*2.28*) $$(s - r)d^T > \left[r - w\frac{d(l_b - l_a)}{d(k_b - k_a)}\right]1 - d^T.$$

Or, to put it more simply, if the time between the firm's action and regulation (*T*) is small enough, then the left hand term in equation (*2.28*) will be greater than the right hand term, and as a result, overcapitalization is worthwhile. If, on the other hand, *T* is significantly large, then in fact the firm will find it worthwhile to remain at point A and not overcapitalize despite the imposition of rate-of-return regulation.[8]

One of the more interesting implications from this regulatory lag relates to research and development by the firm. Baumol and Klevorick (1970) use the regulatory lag model articulated by Bailey and Coleman (1971), to suggest that one possible way for the regulated firm to improve profits *before regulation*, is by research and innovation. After regulation (time *T*), however, prices are readjusted and this readjustment of course takes into account the firm's improved technology, such that the process must begin all over again.

In support of this notion, Baumol and Klevorick speculate that the key in terms of research and innovation, is the length of time before the regulation. They state:

> Suppose the equilibrium stock of knowledge increases with the length of the regulatory lag, because the firm undertakes more research the longer are the periods during which it enjoys the temporary benefit of an excess earnings flow. Then the total discounted present value of the benefits generated by the regulated

firm will rise when the length of the regulatory lag increases
(Baumol and Klevorick 1970, 185).

Although, they do not present a formal model of this process, their
proposition is intuitively appealing.[9]

2.3.2 Non-Profit Maximization

Critical to the Averch-Johnson model is the assumption that firms
maximize profits. Elizabeth Bailey and John Malone (1970) argue that
if firms maximize something other than profits, the regulated firm may
not overcapitalize.[10] Specifically, they find that if the regulated firm
maximizes either revenue or quantity, then it will tend to
undercapitalize.[11] Similarly, Zajac (1970) investigates the case where
the firm maximizes rate of return to equity, and finds that here again the
firm undercapitalizes.

In modeling a rate of return regulated firm, Bailey and Malone
accept many of the same assumptions as Averch and Johnson. The
model is static as before, depreciation and regulatory lags are ignored,
the allowed rate of return set by the regulators (s) is assumed to always
be greater than the firm's cost of capital, and the firm is assumed to
produce only a single product.[12]

Hence, the problem for the profit maximizing monopolist is to
maximize profits subject to the regulatory constraint, where profits for
the unregulated firm is defined as

$$\Pi = R(q(k,l)) - rk - wl ,$$

and where R is sales revenue. Consequently, the problem for the profit
maximizing regulated firm can be stated as the Lagrangian:

$$(2.29) \qquad L(k,l,\lambda) = R(q(k,l)) - rk - wl - \lambda[R(q(k,l)) - sk - wl].$$

Note, that this expression is essentially similar to the Lagrangian
expression defined by Averch and Johnson (see equation 2.7), except
that revenue (R) is left in generality, and the regulatory constraint itself
is no longer an inequality.[13]

As a result, the constrained first order conditions are
straightforward, and as before, it can be shown that the firm does not
cost minimize because:

$$\frac{dl}{dk} = \frac{r}{w} - \frac{\lambda}{(1-\lambda)} \cdot \frac{(s-r)}{w}.$$

Since it is assumed that the allowed rate of return must be greater than the cost of capital ($s > i$), then

$$\frac{\partial l}{\partial k} < \frac{r}{w}$$

for $1 \geq \lambda > 0$. In short, the profit maximizing firm overcapitalizes and does not cost minimize.

If, on the other hand, the firm maximizes revenue instead of profit, then the Lagrangian expression can be restated as:

$$(2.30) \quad L(k,l,\lambda) = R(q(k,l)) - \lambda[R(q(k,l)) - sk - wl].$$

Although nothing has changed with respect to the regulatory constraint, the first term in the Lagrangian is now simply revenue (R). Hence, the first order conditions are:[14]

$$(1-\lambda)\frac{dR}{dq}\frac{\partial q}{\partial k} + \lambda s = 0,$$

$$(1-\lambda)\frac{dR}{dq}\frac{\partial q}{\partial l} + \lambda w = 0,$$

and

$$R(q(k,l)) - sk - wl = 0.$$

Rearranging terms shows that

$$dk = \frac{(1-\lambda)\frac{dR}{dq} \cdot \partial q}{-s\lambda},$$

and

$$dl = \frac{(1-\lambda)\frac{dR}{dq} \cdot \partial q}{-w\lambda}.$$

Hence,

$$(2.31) \quad \frac{dl}{dk} = \frac{(1-\lambda)\dfrac{dR}{dq} \cdot \partial q}{-w\lambda} \frac{-s\lambda}{(1-\lambda)\dfrac{dR}{dq} \cdot \partial q} = \frac{s}{w}.$$

Consequently, the revenue maximizing firm will not cost minimize. Cost minimization occurs only when the marginal rate of substitution of capital for labor is exactly equal to the ratio if the two factor input costs (in this case, the ratio of r to s). Moreover, since it is assumed that $s > r$, then

$$\frac{dl}{dk} > \frac{r}{w}.$$

In short, the revenue maximizing firm will tend to undercapitalize.

This scenario is depicted in figure 2.9. At point E the firm cost minimizes because the slope of the isoquant (dl / dk) is equal to the ratio of the two factor input costs (r / w). At point R_o, however, the firm is not cost minimizing because the slope of the isoquant is greater than the slope of the ratio of the two factor input prices. Hence, the revenue maximizing regulated firm, portrayed in equation (2.32), will use labor beyond the cost minimizing amount and capital below the cost minimizing amount.

If the monopolist maximizes output instead of profits or revenue, then the Lagrangian expression can be defined as:

$$(2.32) \quad L(k,l,\lambda) = q(k,l)) - \lambda\left[q(k,l) - \frac{sk + wl}{P(q)}\right].$$

Hence, the first order conditions are:[15]

$$(1-\lambda)\frac{\partial q}{\partial k} + \lambda\frac{s}{P(q)} = 0,$$

$$(1-\lambda)\frac{\partial q}{\partial l} + \lambda\frac{w}{P(q)} = 0,$$

and

$$q(k,l) - \frac{sk + wl}{P(q)} = 0.$$

Rearranging shows that

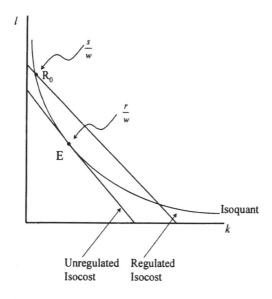

Figure 2.9 Non-Profit Maximization

$$dk = \frac{(1-\lambda)P \cdot \partial q}{-s\lambda},$$

and

$$dl = \frac{(1-\lambda)P \cdot \partial q}{-w\lambda}.$$

Hence,

$$(2.33) \qquad \frac{dl}{dk} = \frac{(1-\lambda)P \cdot \partial q}{-w\lambda} \frac{-s\lambda}{(1-\lambda)P \cdot \partial q} = \frac{s}{w}.$$

Thus, the quantity maximizing firm that is subject to rate of return regulation will not cost minimize. Instead, this firm will tend to undercapitalize because

$$\frac{dl}{dk} > \frac{r}{w}$$

which follows from equation (*2.33*) and because it is assumed that $s > r$. This outcome, which exactly mimics the outcome of the revenue maximizing firm, is also portrayed in figure 2.9. At point E, the firm cost minimizes because the slope of the isoquant (dl / dk) is equal to the ratio of the two factor input costs r and w, but at point R_o the firm is not cost minimizing because $s > r$. Point R_o follows from equation (*2.33*), and demonstrates that the output maximizing regulated firm will use labor beyond the cost minimizing amount, and use capital below the cost minimizing amount.

Zajac (1970) considers the case where the firm maximizes net return per dollar of stockholder investment instead of profit. Zajac makes all the usual assumptions: time, regulatory lags, and depreciation are ignored, and the firm is assumed to produce one output using two factors of production (k and l), the cost of capital is defined as r, and the cost of labor is defined as w. Accordingly, rate of return to stockholder equity (r_e) is defined as:

$$(2.34) \quad r_e = \frac{P(q(k,l))q(k,l) - wl - i_d f_d k}{f_e k}$$

$$= \frac{1}{f_e}\left[\frac{P(q(k,l))q(k,l) - wl}{k}\right] - \frac{i_d f_d k}{f_e}$$

where i_d is the bond rate, $f_d k_i$ is the amount of debt capital, and $f_e k_i$ is the amount of equity capital. Thus, the problem for the firm is maximize (*2.34*) subject to the rate of return regulatory constraint:

$$(2.35) \quad s \geq \frac{P(q(k,l))q(k,l) - wl}{k}.$$

Zajac then notes that the bracketed term in equation (*2.34*) is the same as the left-hand term of the regulatory constraint (*2.21*). As a result, it must be the case that return on equity (r_e) is maximized at the maximum allowed rate of return (s). Hence, the firm is concerned only with operating with input combination (k and l) that result in the largest possible allowed rate of return. Since all points on the regulatory constraint pictured in figure 2.10 represent points where s is maximized, then the firm is indifferent between all such input combinations.

Consequently, because of this indifference, the firm is not driven to a unique capital and labor mix, and must therefore choose where to

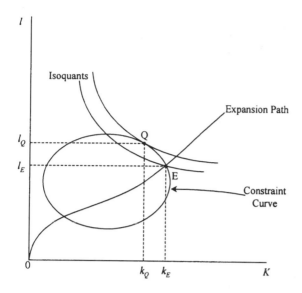

Figure 2.10 Maximizing Return on Stockholder's Equity

operate based on some additional criteria. If, for example, the firm wishes to minimize cost, then it would operate at point E in figure 2.10. This point of course entails no overuse of capital or labor. If, on the other hand, the firm wishes to maximize output, then it would choose point Q given the isoquants shown. Interestingly, in this particular case, capital is in fact underused and thus the effect of rate of return regulation is the opposite of the Averch-Johnson effect. Other cases would obviously result in different outcomes.[16]

2.3.3 The Cost of Capital and the Allowed Rate-of-Return

Averch and Johnson assume that the allowed rate-of-return is greater than the cost of capital ($s > r$). Indeed this is an important assumption of their model, as well as the subsequent extensions of the Averch-Johnson model. On this subject, Baumol and Klevorick argue that:

> This assertion is not very easy to test empirically, since the cost of capital presumably varies from firm to firm, depending at least in part on the degree of risk incurred by anyone who provides capital

to it, and we do not really know how to measure that cost with any
high degree of precision... it is at least arguable that the regulator
often attempts to set the "fair rate of return" at a level as close to
the cost of capital as he can determine (Baumol and Klevorick
1970, 173).

Moreover, Baumol and Klevorick note that regulators generally seek to
equate the allowed rate of return to the cost of capital ($s = r$). If this is
accomplished, then the Averch-Johnson effect becomes untenable and
the firm has no incentive to operate at any particular capital-labor ratio.[17]
    Baumol and Klevorick demonstrate this by noting that in the
relevant first order condition from the Averch-Johnson Lagrangian,
(equation *2.8*), is:

$$r \geq \left[1 - \lambda\right]\left[P + q\frac{dP}{dq}\right]\frac{\partial q}{\partial k} + \lambda s .$$

Assuming $\lambda > 0$, then $\lambda = 1$ if and only if $s = r$. Hence, if $s = r$, then
only $\lambda = 1$ can satisfy this condition.   Given this, then the profit
constraint (see equation *2.5*), for the monopolist can be stated as :

$$r = s \geq \frac{P(q(k,l))q(k,l) - wl}{k} .$$

As a result, any values of k and l that satisfy this constraint will also
satisfy the first order conditions.
    To see this geometrically, recall that the profit constraint defined in
figure 2.5 hinges on the l-axis and swings upward along the k-axis. The
slope of this constraint plane is determined by the difference between the
allowed rate of return and the cost of capital ($s - r$). If s exactly equals r,
then the constraint plane is flat (zero slope) and connects the entire
portion of the l-axis with the k-axis. Thus when $s = r$, regulation offers
the firm no possibility of improving profit, so the firm has no motivation
to use to much capital, or to use too little capital. In fact, the firm has no
motivation to use any particular capital-labor mix, since with all possible
combinations the profit will be the same: zero.

2.3.4 Induced Technical Change

The lone attempt to investigate the impact of rate of return regulation on the technological innovation of firms is that of V. Kerry Smith (1974), with a subsequent discussion found in Smith (1975). Subsequent works have all been empirical attempts to estimate Smith's propositions (see for example Macauley 1986; Nelson 1984; and Granderson 1999).[18]

The assumptions of Smith's model mimic the assumptions of Bailey and Malone (1970), not the relatively more strict assumptions of Averch and Johnson (1962). Specifically, Smith ignores depreciation, time, and any regulatory lags and assumes that the firm produces one output that requires two factors for nonzero production (k_i and l_i), that the cost of labor is the wage rate (w), that the cost of capital is the interest rates (r), and that the regulatory constraint is completely effective such that the regulatory is not an inequality. Hence profit can be generally defined as:

$$\Pi = R(q(k,l)) - rk - wl \,,$$

subject to a fair rate of return on investment which is defined as:

$$[s - r]k \,.$$

As in Bailey and Malone (1970), R is sales revenue for the monopolist, q is output, k is capital investment, l is labor, r is interest rate and w is wage rates.

Moreover, Smith also assumes that technology affects the firm by either augmenting labor or capital, such that the production function for the firm is defined as:

$$q = q(a_1 l, a_2 k)$$

where a_1 is labor augmenting technology, and a_2 is capital augmenting technology. The amount of effort put towards innovation is given by T, which is assumed to be priced at Θ dollars per unit. It is further assumed that the shape of the T frontier is concave as a result of diminishing returns to research and development.

Consequently, the problem for the profit maximizing regulated firm can be stated by the Lagrangian:[19]

$$(2.36) \quad L(k, l, a_1, a_2 \lambda) = R(q(a_1 l, a_2 k)) - rk - wl - \Theta T(a_1, a_2)$$
$$- \lambda \big[R(q(a_1 l, a_2 k)) - sk - wl - \Theta T(a_1, a_2) \big].$$

Accepting that this constraint is binding, then the first order conditions are:

$$(2.37) \quad (1-\lambda)\left[a_2 \frac{dR}{dq} \frac{\partial q}{\partial(a_2 k)}\right] - r + \lambda s = 0,$$

$$(2.38) \quad (1-\lambda)\left[a_1 \frac{dR}{dq} \frac{\partial q}{\partial(a_1 l)} - w\right] = 0,$$

$$(2.39) \quad (1-\lambda)\left[l \frac{dR}{dq} \frac{\partial q}{\partial(a_1 l)} - \Theta \frac{\partial T}{\partial a_1}\right] = 0,$$

$$(2.40) \quad (1-\lambda)\left[k \frac{dR}{dq} \frac{\partial q}{\partial(a_2 k)} - \Theta \frac{\partial T}{\partial a_2}\right] = 0,$$

and

$$(2.41) \quad R(q(a_1 l, a_2 k)) - sk - wl - \Theta T(a_1, a_2) = 0.$$

By rearranging equations (2.37) and (2.38), and then combining these two, the marginal rate of substitution between labor and capital can be found. Thus, equation (2.37) can be rearranged as:

$$\frac{\partial q}{\partial(a_2 k)} a_2 = \frac{(r-\lambda)s}{1-\lambda} \cdot \frac{1}{\frac{dR}{dq}}.$$

Likewise, equation (2.38) can be rearranged as:

$$\frac{\partial q}{\partial(a_1 l)} a_1 = \frac{w}{1} \cdot \frac{1}{\frac{dR}{dq}}.$$

Combining these rearrangements defines the marginal rate of substitution between capital and labor:

$$(2.42) \quad \frac{a_2}{a_1} \cdot \frac{\partial q/\partial(a_2 k)}{\partial q/\partial(a_1 l)} = \frac{r-\lambda s}{w(1-\lambda)}.$$

Note that this expression of the marginal rate of substitution of capital for labor is essentially identical to the Averch-Johnson finding (see equation 2.16). As a result, if no regulation is imposed on the firm ($\lambda = 0$), this statement of the marginal rate of substitution thus becomes equal to the ratio of the wage cost to the cost of capital, as would be expected.

If some degree of regulation is imposed on the firm ($\lambda > 0$), then the firm overuses capital relative to labor.

By rearranging and combining equations (2.39) and (2.40) the marginal rate of substitution between the two technology factors (a_1 and a_2), can be revealed. Equation (2.39) can be restated as:

$$\frac{\partial T}{\partial a_1} = \frac{l}{\Theta} \cdot \frac{dR}{dq} \cdot \frac{\partial q}{\partial(a_1 l)},$$

and equation (2.40) as:

$$\frac{\partial T}{\partial a_2} = \frac{k}{\Theta} \cdot \frac{dR}{dq} \cdot \frac{\partial q}{\partial(a_2 k)}.$$

Combining these two restatements reveals:

$$(2.43) \quad \frac{\partial T/\partial a_1}{\partial T/\partial a_2} = \left(\frac{dR}{dq} \frac{l}{\Theta} \right) \frac{\partial q}{\partial(a_1 l)} \cdot \left(\frac{1}{\dfrac{dR}{dq} \dfrac{k}{\Theta}} \right) \frac{\partial(a_2 k)}{\partial q}$$

$$= \frac{l}{k} \cdot \frac{\partial q/\partial(a_1 l)}{\partial q/\partial(a_2 k)}$$

which is the marginal rate of substitution between a_1 and a_2.

If the firm solves both equations (2.42) and (2.43) simultaneously, as is implicitly assumed by Smith, then the regulated firm will overcapitalize and will select technology that augments labor more than capital. To see this, equation (2.42) must be rearranged to the form:

$$\frac{a_2}{a_1} = \frac{\partial q/\partial(a_1 l)}{\partial q/\partial(a_2 k)} \cdot \frac{r - \lambda s}{w(1 - \lambda)},$$

and equation (2.43) must be rearranged to:

$$\frac{\partial q/\partial(a_1 l)}{\partial q/\partial(a_2 k)} = \frac{\partial T/\partial a_1}{\partial T/\partial a_2} \cdot \frac{k}{l}.$$

After substituting, the ratio of the two factor technologies can be stated as:

$$(2.44) \quad \frac{a_2}{a_1} = \frac{\partial T/\partial a_1}{\partial T/\partial a_2} \cdot \frac{k}{l} \cdot \frac{r - \lambda s}{w(1 - \lambda)}$$

$$= \frac{\partial T/\partial a_1}{\partial T/\partial a_2} \cdot \frac{1 - \lambda \frac{s}{r}}{1 - \lambda} \cdot \frac{rk}{wl}$$

This is identical to equation (6) in Smith.[20]

From equation (2.44), it becomes clear, as Smith contends, that since $0 < \lambda < 1$, $s > r$, and $\lambda s < r$, the regulated firm will select technology that augments labor more than capital. This follows because, given these assumptions,

$$\frac{a_2}{a_1} < \frac{\partial T/\partial a_1}{\partial T/\partial a_2} \cdot \frac{rk}{wl}$$

for the regulated firm. Clearly

$$\left(1 - \lambda \frac{s}{r}\right) < (1 - \lambda),$$

because $\lambda s < r$ and $0 < \lambda < 1$. As a result it must be the case that for all regulated firms

$$\frac{1 - \lambda \frac{s}{r}}{1 - \lambda} < 1,$$

and consequently

$$\frac{a_2}{a_1} < \frac{\partial T/\partial a_1}{\partial T/\partial a_2} \cdot \frac{rk}{wl}.$$

If, on the other hand, the firm is unregulated ($\lambda = 0$), then equation (2.44) can be stated as:

$$\frac{a_2}{a_1} = \frac{\partial T/\partial a_1}{\partial T/\partial a_2} \cdot \frac{rk}{wl},$$

in which the choice of capital and labor technologies is efficient.

Consequently, Smith argues that rate-of-return regulation has two important effects on the firm: first, as a result of regulation, the regulated firm will overuse capital relative to labor, and second, the regulated firm will select technology which augments labor more than capital. In Smith's words, "a profit maximizing firm subject to a fair return on investment regulation will overcapitalize and select those technical changes which will allow it to continue to do so – namely labor augmenting innovations" (Smith 1974, 630).

Notes

[1] Because of this similarity in conclusion with that of the Averch and Johnson conclusion, Alfred Kahn (1988) calls the Averch-Johnson effect, the Averch-Johnson-Wellisz effect (see Kahn 1988, volume II, 49-59). Although this seems a bit extreme given the very narrow focus of Wellisz and the more general approach and conclusion of Averch and Johnson, the point of similarity is clear.

[2] Alternatively, this could be altered to the Cournot case where $P(Q(\Sigma q_i))$ and $dQ/dq_i = 1$.

[3] In fact, time is not at all considered in any of their modeling. They prefer the static model because, as they content, to construct a dynamic model "would complicate the results without contributing much additional insight into the behavior of the firm" (p. 1054, Averch and Johnson, 1962).

[4] Zajac argues that multiple peaks with local constrained profit maxima are certainly possible but only tend to "obscure the economic issues" (Zajac 1970, 118). Hence, Zajac explicitly ignore such contingencies, as Averch and Johnson do implicitly.

[5] Note also that profit is not constant along the constraint curve.

[6] Train, for example, says "their model and conclusions have been questioned from a number of perspectives, and, in fact, some errors in their logic have been discovered (though these errors do not affect their essential conclusions)" (Train 1991, 19).

[7] The exception is when the rate of return is set exactly equal to the cost of capital (see section 2.3.3).

[8] Davis (1973) makes a roughly similar conclusion. Davis uses a dynamic model with a regulatory lag in the price adjustment mechanism, and finds that the firm still tends to overcapitalize, although this overcapitalization is less than the simple Averch-Johnson model, but more than the efficient point.

[9] Klevorick (1973) does attempt to incorporate the process of technical change into an elaborate model of the a firm subject to discrete rate of return regulation, although these results shed no additional light on this issue. Indeed, this rather complex attempt by Klevorick can best be described as obscuring rather than invalidating the argument.

[10] Baumol and Klevorick (1970) also note this problem, although their discussion is brief and based largely on the findings contained in Bailey and Malone (1970).

[11] Bailey and Malone also investigate the case when the firm maximizes its return on investment, but are unable to determine the outcome.

[12] One distinction is that Bailey and Malone use interest rates as the cost of capital (r).

[13] This is an important simplifying assumption that most subsequent work on the Averch-Johnson replicates. If the constraint is stated as an inequality, then the Khan-Tucker theorem must be used to get the first order conditions. But by assuming that regulation is

completely effective, then the constraint is no longer an inequality (see also Takayama 1969).

[14] Unfortunately, the first order conditions and rearranging that follows, is not reported in the article by Bailey and Malone (1970). These steps are only reasonably inferred from their Lagrangian setup and the final conclusion.

[15] As noted before, the first order conditions and the rearranging that follows, is not reported in Bailey and Malone (1970).

[16] More critically, it should be noted that this example in part depends on the nature of the expansion path. If instead the expansion path happens to pass through the constraint curve to the left of the output maximizing level, then Q would entail overusing capital beyond the efficient level, somewhat more like the Averch-Johnson effect.

[17] The remaining possibility is of course if $s < r$. In this case, as both Averch and Johnson (1962) and Baumol and Klevorick (1970) point out, the firm will as a result lose money and in the long run be forced to shut down.

[18] See chapter 8, section 8.1 for a detailed discussion of the empirical literature.

[19] Smith (1974, 1975) does not report this Lagrangian, nor any of the first order conditions, nor of the manipulations which generate the final results, thus all of the following must be inferred. Instead, Smith simply states "the first-order conditions are essentially similar to those of Bailey and Malone (1970) with the addition of conditions for a_1 and a_2, and our additional constraint" (Smith 1974, 627).

[20] To reach equation *(2.44)*, of course it must be assumed that equations *(2.42)* and *(2.43)* can be solved simultaneous by the firm, which Smith implicitly accepts. Okuguchi (1975), however, shows with a constant elasticity production function, that this may not always be the case, and upon such occasions, Smith's conclusions are not tenable.

3 Augmented Averch-Johnson

Building upon the model articulated by Averch and Johnson (1962), this chapter considers the case of a firm with market power that produces a single product, using three input factors, capital and labor as in Averch-Johnson, and a third factor, research and development. We define:

$P(Q)$ = price of output Q
q_i = output of the ith firm
c_i = total costs of the ith firm
r = return on capital investment
s = allowed regulatory return on capital investment
k_i = capital of the ith firm
w = wage rate
l_i = labor of the ith firm
γ = R&D rate
x_i = R&D investment of the ith firm.

Following the presentation of the three-input model, a more specific augmentation of the Averch-Johnson model is discussed in which technology is allowed to augment labor and capital separately, and an additional parameter catches the rate of research spillovers. An appendix contains a brief comparison of rate of return regulation using a neoclassical production function and an endogenous production function.

3.1 Augmented Averch-Johnson

Let total costs (c) be a function of research and development (x), capital (k), and labor (l), such that the costs of production for firm i are:[1]

$$(3.1) \qquad c_i(x_i, k_i, l_i) = rk_i + wl_i + \gamma x_i.$$

Let production be a function of each of the three inputs, such that all three are needed for nonzero production, and let the R&D investments by other firms not affect firm i's production.[2] Hence:

(3.2) $q_i = q_i(x_i, k_i, l_i),$

where

$\qquad x_i > 0, \quad k_i > 0, \quad l_i > 0,$

and

$$\frac{\partial q_i}{\partial x_i} > 0, \qquad \frac{\partial q_i}{\partial k_i} > 0, \qquad \frac{\partial q_i}{\partial l_i} > 0.$$

Thus, profit for the unregulated firm is defined as:

(3.3) $\Pi_i = P(Q(q_i(x_i, k_i, l_i)))q_i(x_i, k_i, l_i) - rk_i - wl_i - \gamma x_i.$

The imposition of rate-of-return regulation is intended to allow firms just sufficient revenue net of operation expenses to compensate for its investment in plant and equipment. Ignoring depreciation costs, then this is equivalent to the requirement that the firm's revenue net of noncapital costs divided by the level of capital investment not be greater than the government imposed maximum rate of return s. Thus,

(3.4) $s \geq \dfrac{P(Q(q_i(x_i, k_i, l_i)))q_i(x_i, k_i, l_i) - wl_i - \gamma x_i}{k_i}.$

This is similar to the specification of the rate of return regulatory constraint as put forth by Averch and Johnson (1962). In fact, the only significant difference is the addition of a third factor of production, x_i. For simplicity, it is further assumed that this regulatory constraint is binding (see for example Takayama 1969, Bailey and Malone 1970, or Smith 1975),[3] thus

(3.5) $P(Q(q_i(x_i, k_i, l_i)))q_i(x_i, k_i, l_i) - sk_i - wl_i - \gamma x_i = 0.$

Consequently, the problem for the regulated firm with profit function (3.3), is to maximize profits subject to the regulatory constraint (3.5):

(3.6) $L(x_i, k_i, l_i, \lambda) = P(Q(q_i(x_i, k_i, l_i)))q_i(x_i, k_i, l_i) - rk_i$

$\qquad\qquad - wl_i - \gamma x_i - \lambda[P(Q(q_i(x_i, k_i, l_i)))q_i(x_i, k_i, l_i) - sk_i$

$\qquad\qquad - wl_i - \gamma x_i],$

where

$\qquad 0 \leq \lambda \leq 1,$

and

$\qquad s > r.$

Assuming Cournot conjectures (hence, $dQ/dq_i = 1$), then the necessary constrained first-order conditions are:

$$(3.7) \qquad 0 = (1 - \lambda)\left[\left(P + q_i \frac{dP}{dq_i}\right)\frac{\partial q_i}{\partial x_i} - \gamma\right],$$

$$(3.8) \qquad 0 = (1 - \lambda)\left(P + q_i \frac{dP}{dq_i}\right)\frac{\partial q_i}{\partial k_i} + s\lambda - r,$$

$$(3.9) \qquad 0 = (1 - \lambda)\left[\left(P + q_i \frac{dP}{dq_i}\right)\frac{\partial q_i}{\partial l_i} - w\right],$$

and

$$(3.10) \qquad 0 = P(Q(q_i(x_i, k_i, l_i)))q_i(x_i, k_i, l_i) - sk_i - wl_i - \gamma x_i.$$

It follows that $\lambda = 0$ indicates the complete absence of effective regulation, and $\lambda = 1$ indicates the complete effectiveness of regulation. Rearranging equation (3.7), we note the marginal product of R&D:

$$MP_x = \frac{\partial q_i}{\partial x_i} = \frac{\gamma(1 - \lambda)}{(1 - \lambda)\left(P + q_i \frac{dP}{dq_i}\right)}.$$

Similarly, by rearranging equations (3.8) and (3.9):

$$MP_k = \frac{\partial q_i}{\partial k_i} = \frac{r - s\lambda}{(1 - \lambda)\left(P + q_i \frac{dP}{dq_i}\right)},$$

and

$$MP_l = \frac{\partial q_i}{\partial l_i} = \frac{w(1 - \lambda)}{(1 - \lambda)\left(P + q_i \frac{dP}{dq_i}\right)}.$$

Further rearranging reveals the marginal rates of substitution:

$$(3.11) \qquad MRS_{kl} = \frac{\partial q_i/\partial k_i}{\partial q_i/\partial l_i} = \frac{r - s\lambda}{w(1 - \lambda)},$$

$$(3.12) \quad MRS_{kx} = \frac{\partial q_i / \partial k_i}{\partial q_i / \partial x_i} = \frac{r - s\lambda}{\gamma(1 - \lambda)},$$

$$(3.13) \quad MRS_{lx} = \frac{\partial q_i / \partial l_i}{\partial q_i / \partial x_i} = \frac{w(1 - \lambda)}{\gamma(1 - \lambda)} = \frac{w}{\gamma}.$$

Proposition 3.1. *In this augmented Averch-Johnson model, as* λ *approaches zero, the marginal rate of substitution of capital for labor becomes equal to the ratio of r to w.*

This follows from equation (*3.11*) which gives the marginal rate of substitution of capital to labor. It implies that if there is no effective rate of return regulation imposed on the firm (that is, if $\lambda = 0$), then the marginal rate of substitution of labor for capital must equal the ratio of the two factor input prices:

$$(3.14) \quad \frac{\partial q_i / \partial k_i}{\partial q_i / \partial l_i} = \frac{r}{w}.$$

Moreover, since $s > r$:

$$\lim_{\lambda \to 0^+} \frac{\lambda}{1 - \lambda} \cdot \frac{s - r}{w} = 0,$$

and as a result

$$\lim_{\lambda \to 0^+} \left(\frac{r}{w} - \frac{\lambda}{1 - \lambda} \cdot \frac{s - r}{w} \right) = \frac{r}{w}.$$

Consequently, as λ approaches zero (the unregulated case) the firm will tend to use an efficient mix of capital and labor.

Proposition 3.2 *In the augmented Averch-Johnson case, if* $0 < \lambda < 1$, *then the marginal rate of substitution of capital for labor will be less than the ratio of r and w.*

This again follows from equation (*3.11*). If regulation is effective ($0 < \lambda < 1$), then

$$\frac{\partial q_i / \partial k_i}{\partial q_i / \partial l_i} = \frac{r - s\lambda}{w(1 - \lambda)} = \frac{r}{w} - \frac{\lambda}{(1 - \lambda)} \frac{(s - r)}{w}$$

where

$$\frac{\lambda}{[1-\lambda]}\frac{[s-r]}{w} > 0,$$

because $s > r$. As a result, it must be the case that

(3.15) $$\frac{\partial q_i/\partial k_i}{\partial q_i/\partial l_i} < \frac{r}{w}.$$

In short, the firm under some degree of effective regulation will overuse capital relative to labor, such that it operates at an output where the marginal rate of substitution of capital for labor will not equal the ratio of r to w. This situation is depicted in figure 3.1. Holding R&D constant, point E in the figure represents unregulated case of equation (3.14), point R on the other hand, represents the case of effective regulation from equation (3.15). Although output is identical at both points, R requires a greater isocost, as is noted in the figure. This is of course the familiar Averch-Johnson result.

Proposition 3.3. *As λ approaches one, the marginal rate of substitution of capital for labor will become infinitely less than the ratio of the factor input prices r and w.*

This follows because:

$$\lim_{\lambda \to 1}\frac{\lambda}{1-\lambda}\cdot\frac{s-r}{w} = \infty,$$

hence,

$$\lim_{\lambda \to 1}\left(\frac{r}{w} - \frac{\lambda}{1-\lambda}\cdot\frac{s-r}{w}\right) = -\infty.$$

That is, as regulation becomes completely effective ($\lambda = 1$ and thus $s = r$), then the marginal rate of substitution of capital for labor becomes infinitely less than the ratio of r to w.

Proposition 3.4   *In the augmented Averch-Johnson model, as λ approaches zero, then the marginal rate of substitution of R&D for capital will equal the ratio of the factor input prices r and γ.*

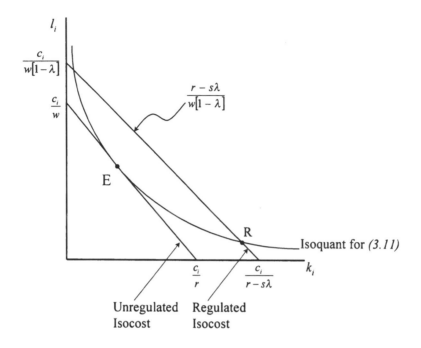

Figure 3.1 The Effects of Regulation on Capital and Labor Inputs

Given that $s > r$ and $0 < \lambda < 1$, then from equation *(3.12)* :

$$\lim_{\lambda \to 0^+} \frac{\lambda}{1-\lambda} \cdot \frac{s-r}{\gamma} = 0 ,$$

hence,

$$\lim_{\lambda \to 0^+} \left(\frac{r}{\gamma} - \frac{\lambda}{1-\lambda} \cdot \frac{s-r}{\gamma} \right) = \frac{r}{\gamma} .$$

As a result, if $\lambda = 0$ then:

(3.16) $\qquad \dfrac{\partial q_i / \partial k_i}{\partial q_i / \partial x_i} = \dfrac{r}{\gamma} .$

Consequently, the unregulated firm will tend to use an efficient mix of capital to R&D.

Proposition 3.5 *If $0 < \lambda < 1$, then the marginal rate of substitution of capital for R&D will be less than the ratio of r and γ.*

This is because when regulation is effective ($0 < \lambda < 1$),

$$\frac{\lambda}{[1 - \lambda]} \frac{[s - r]}{\gamma} > 0$$

because $s > r$. Using this in equation (*3.12*), then it must be the case that

$$(3.17) \qquad \frac{\partial q_i / \partial k_i}{\partial q_i / \partial x_i} < \frac{r}{\gamma}.$$

Consequently, the firm under some degree of effective regulation will overuse capital relative to R&D, such that it operates at an output where the marginal rate of substitution of capital for R&D will not equal the ratio of r to γ. This is the situation depicted in Figure 3.2 where, holding labor constant, the regulated firm substitutes capital for research and development. Point E in the figure represents the situation of the unregulated firm described in equation (*3.16*), whereas point R represents the regulated firm from equation (*3.17*).

Proposition 3.6. *As λ approaches one, the marginal rate of substitution of capital for R&D will become infinitely less than the ratio of r and γ.*

This follows from equation (*3.12*) because:

$$\underset{\lambda \to 1}{Lim} \frac{\lambda}{1 - \lambda} \cdot \frac{s - r}{\gamma} = \infty,$$

hence,

$$\underset{\lambda \to 1}{Lim} \left(\frac{r}{\gamma} - \frac{\lambda}{1 - \lambda} \cdot \frac{s - r}{\gamma} \right) = -\infty.$$

That is, as regulation (λ) becomes completely effective ($s = r$), the marginal rate of substitution of capital for R&D becomes infinitely less than the ratio of the two factor prices r and γ.

Proposition 3.7. *In the augmented Averch-Johnson case, regulation will have no impact on the marginal substitution of labor for R&D, and the*

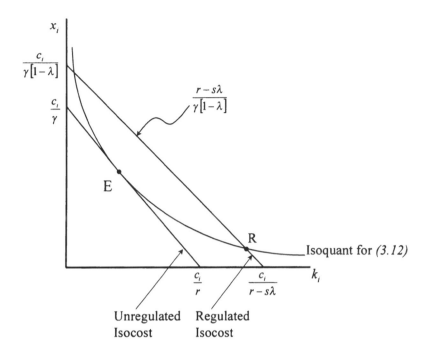

Figure 3.2 The Effects of Regulation on Capital and R&D Inputs

firm will tend to operate where this marginal rate of substitution exactly equals the ratio of w to γ.

This follows directly from equation (*3.13*) which states:

$$\frac{\partial q_i / \partial l_i}{\partial q_i / \partial x_i} = \frac{w}{\gamma}.$$

Hence, the firm tends to use an efficient mix of labor and R&D regardless of the degree or the effectiveness of government regulation (λ).   Consequently, when effective regulation occurs, capital is substituted for both labor and R&D, yet labor and R&D are not substituted for one another. This scenario is depicted in figure 3.3. The inputs of x_i and l_i are such that their costs, γ and w respectively, are equal to their marginal value products eve:: under effective regulation. The

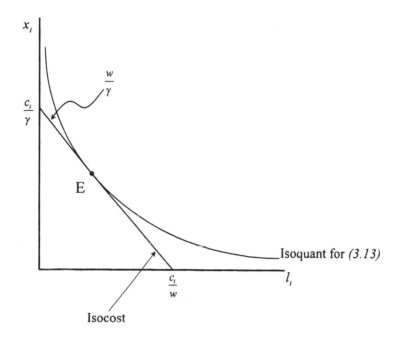

Figure 3.3 The Effects of Regulation on Labor and R&D Inputs

input of k_i, however, is such that its use is expanded beyond the point where its cost r would be equal to its marginal value product.

Table 3.1 presents a brief summary of the propositions from this section. Most importantly, in the augmented Averch-Johnson case, the unregulated firm tends to use its capital, labor, and R&D inputs efficiently, while the regulated firm tends to overuse capital relative to labor and R&D. With respect to the marginal substitution of capital for labor, these conclusions are identical to the original conclusions of Averch and Johnson.

In this model research and development is treated as simply a third factor input, indistinguishable in the production function from capital or labor. Although this does have precedence in the growth theory literature (see for example, Uzawa 1965, Jones and Manuelli 1990, and Rebelo 1991), it is useful to articulate precise avenues by which technology can affect the firm's production. To this end, the following

Table 3.1 Regulatory Implications from the Augmented Averch-Johnson Model

Firm's Profit Maximizing Objective	*Unregulated Case* $(\lambda \to 0^{+})$	*Regulated Case* $(\lambda \to 1)$
$\dfrac{\partial l_i}{\partial k_i} = \dfrac{r}{w} - \dfrac{\lambda}{(1-\lambda)}\dfrac{(s-r)}{w}$	Efficient mix of capital and labor	Overuse capital relative to labor
$\dfrac{\partial x_i}{\partial k_i} = \dfrac{r}{\gamma} - \dfrac{\lambda}{(1-\lambda)}\dfrac{(s-r)}{\gamma}$	Efficient mix of R&D and capital	Overuse capital relative to R&D
$\dfrac{\partial x_i}{\partial l_i} = \dfrac{w}{\gamma}$	Efficient mix of R&D and labor	Efficient mix of R&D and labor

section will investigate the narrower case in which technology is separated into labor-augmenting and capital-augmenting technology.

3.2   Extension on Factor Augmentation with R&D Spillovers

In this section it is assumed that the augmentation of the factors of production by technology is defined specifically, such that production for the ith firm is:

$$q_i = f\left(k_i \chi^k, l_i \chi^l\right),$$

where

$$\chi^k = x_i^k + \sigma \sum_{i \neq j} x_j^k$$

and

$$\chi^l = x_i^l + \sigma \sum_{i \neq j} x_j^l .$$

Accordingly, χ^k is defined as the aggregate stock of capital augmenting R&D technology, and is composed of both firm i's capital augmenting technology $\left(x_i^k\right)$, and the R&D technical spillover from other firms $\left(\sigma \sum_{i \neq j} x_j^k\right)$. Likewise, χ^l is defined as the aggregate stock of labor augmenting R&D technology, and is composed of both firm i's labor augmenting technology $\left(x_i^l\right)$, and the labor augmenting technical spillover from other firms $\left(\sigma \sum_{i \neq j} x_j^l\right)$. The R&D spillover rate is defined by the parameter σ, where $0 \leq \sigma \leq 1$, with $\sigma = 0$, indicating no research spillovers, and $\sigma = 1$ indicating full research spillovers (for similar specifications see for example, Yi 1996, D'Aspremont and Jacquemin 1988, and Poyago-Theotoky 1995).

Hence, then the problem for the regulated firm is to maximize profits subject to the rate-of-return regulatory constraint:

$$(3.18) \quad L(k_i, l_i, x_i^k, x_i^l \lambda) = P(Q(f(k_i \chi^k, l_i \chi^l))) f(k_i \chi^k, l_i \chi^l)$$
$$- rk_i - wl_i - \gamma^k x_i^k - \gamma^l x_i^l - \lambda \big[P(Q(f(k_i \chi^k, l_i \chi^l)))$$
$$\cdot f(k_i \chi^k, l_i \chi^l) - sk_i - wl_i - \gamma^k x_i^k - \gamma^l x_i^l \big],$$

where

$$0 \leq \lambda \leq 1,$$
$$\lambda s < r,^4$$
$$s > r,$$
$$x_i^k(0) = 1, \quad x_i^k \geq 1,$$

and

$$x_i^l(0) = 1, \quad x_i^l \geq 1.$$

If we assume Cournot conjectures ($dQ/dq_i = 1$), then the necessary constrained first-order conditions are:

$$(3.19) \quad 0 = (1 - \lambda)\left(P + f\,\frac{dP}{dq_i}\right)\left(x_i^k + \sigma\sum_j x_j^k\right)\frac{\partial f}{\partial(k_i \chi^j)} - r + s\lambda,$$

$$(3.20) \quad 0 = (1 - \lambda)\left[\left(P + f\,\frac{dP}{dq_i}\right)\left(x_i^l + \sigma\sum_j x_j^l\right)\frac{\partial f}{\partial(l_i \chi^l)} - w\right],$$

$$(3.21) \quad 0 = (1 - \lambda)\left[\left(P + f\,\frac{dP}{dq_i}\right)k_i\,\frac{\partial f}{\partial(k_i \chi^k)} - \gamma^k\right],$$

$$(3.22) \quad 0 = (1 - \lambda)\left[\left(P + f\,\frac{dP}{dq_i}\right)l_i\,\frac{\partial f}{\partial(l_i \chi^l)} - \gamma^l\right],$$

and

$$(3.23) \quad 0 = P(Q(f(k_i\chi^k, l_i\chi^l)))f(k_i\chi^k, l_i\chi^l) - rk_i - wl_i - \gamma^k x_i^k$$
$$- \gamma^l x_i^l .$$

These can be restated as marginal products:

$$(3.24) \quad MP_k = \frac{\partial q_i}{\partial k_i} = \left(x_i^k + \sigma\sum_j x_j^k\right)\frac{\partial f}{\partial(k_i\chi^k)}$$

$$= \frac{r - s\lambda}{(1 - \lambda)\left(P + f\,\dfrac{dP}{dq_i}\right)},$$

$$(3.25) \quad MP_l = \frac{\partial q_i}{\partial l_i} = \left(x_i^l + \sigma\sum_j x_j^l\right)\frac{\partial f}{\partial(l_i\chi^l)}$$

$$= \frac{w(1 - \lambda)}{(1 - \lambda)\left(P + f\,\dfrac{dP}{dq_i}\right)},$$

$$(3.26) \quad MP_{x^k} = \frac{\partial q_i}{\partial x_i^k} = k_i\,\frac{\partial f}{\partial(k_i\chi^k)} = \frac{\gamma^k(1 - \lambda)}{(1 - \lambda)\left(P + f\,\dfrac{dP}{dq_i}\right)},$$

and

$$(3.27) \quad MP_{x^l} = \frac{\partial q_i}{\partial l_i} = l_i \frac{\partial f}{\partial (l_i x^l)} = \frac{\gamma^l(1-\lambda)}{(1-\lambda)\left(P + f\dfrac{dP}{dq_i}\right)}.$$

These are combined to reveal the marginal rates of substitution:

$$(3.28) \quad MRS_{kl} = \frac{MP_k}{MP_l} = \frac{r - s\lambda}{w(1-\lambda)} = \frac{r}{w} - \frac{\lambda}{(1-\lambda)}\frac{(s-r)}{w},$$

$$(3.29) \quad MRS_{kx^k} = \frac{MP_k}{MP_{x^k}} = \frac{r}{\gamma^k} - \frac{\lambda}{(1-\lambda)}\frac{(s-r)}{\gamma^k},$$

$$(3.30) \quad MRS_{kx^k} = \frac{MP_k}{MP_{x^l}} = \frac{r}{\gamma^l} - \frac{\lambda}{(1-\lambda)}\frac{(s-r)}{\gamma^l},$$

$$(3.31) \quad MRS_{lx^k} = \frac{MP_l}{MP_{x^k}} = \frac{w}{\gamma^k},$$

and

$$(3.32) \quad MRS_{lx^l} = \frac{MP_l}{MP_{x^l}} = \frac{w}{\gamma^l}.$$

The implications from these marginal relationships are similar to the augmented Averch-Johnson case presented in section 3.1. Indeed, equation (3.28) which states the marginal rate of substitution of capital for labor, is identical to (3.11). Both equations indicate that the regulated firm uses more capital relative to labor than the unregulated firm. Equations (3.29) and (3.30) state the marginal substitution of capital for capital augmenting technology and labor augmenting technology respectively, and are similar to equation (3.12) in the augmented Averch-Johnson model. The implication from these is that the regulated firm will use more capital relative to capital augmenting technology and labor augmenting technology than the unregulated firm. Equations (3.31) and (3.32) are similar to equation (3.13) and indicate that regulation has no impact on the marginal substitution of labor for either capital augmenting R&D or labor augmenting R&D. Consequently, the unregulated firm tends to use its capital, labor and technology efficiently, while the regulated firm tends to overuse capital relative to labor and technology. Table 3.2 presents a brief summary of these marginal relationships.

Equations (*3.24*) and (*3.25*) can be rearranged as statements of capital augmenting R&D and labor augmenting R&D:

$$\chi^k = \frac{r - s\lambda}{(1-\lambda)\left(P + x_i f \dfrac{dP}{dq_i}\right)\dfrac{\partial f}{\partial(k_i \chi^k)}}$$

and

$$\chi^l = \frac{w(1-\lambda)}{(1-\lambda)\left(P + f \dfrac{dP}{dq_i}\right)\dfrac{\partial f}{\partial(l_i \chi^l)}}.$$

Combining these reveals:

$$\frac{\chi^k}{\chi^l} = \frac{r - s\lambda}{(1-\lambda)\left(P + f \dfrac{dP}{dq_i}\right)\dfrac{\partial f}{\partial(k_i \chi^k)}} \cdot \frac{(1-\lambda)\left(P + f \dfrac{dP}{dq_i}\right)\dfrac{\partial f}{\partial(l_i \chi^l)}}{w(1-\lambda)}$$

$$= \frac{r - s\lambda}{w(1-\lambda)} \cdot \frac{\dfrac{\partial f}{\partial(l_i \chi^l)}}{\dfrac{\partial f}{\partial(k_i \chi^k)}},$$

or

$$(3.33) \qquad \frac{x_i^k}{x_i^l} = \frac{r - s\lambda}{w(1-\lambda)} \cdot \frac{\dfrac{\partial f}{\partial(l_i \chi^l)}}{\dfrac{\partial f}{\partial(k_i \chi^k)}} - \frac{\sigma \sum x_j^l}{\sigma \sum x_j^k}.$$

Equations (*3.21*) and (*3.22*) can be rearranged to:

$$\frac{\partial f}{\partial(k_i \chi^k)} = \frac{\gamma^k (1-\lambda)}{(1-\lambda)\left(P + f \dfrac{dP}{dq_i}\right)k_i}$$

Table 3.2 Implications from the Factor Augmented Averch-Johnson Model

Firm's Profit Maximizing Objective	*Unregulated Case* $(\lambda \to 0^{+})$	*Regulated Case* $(\lambda \to 1)$
$MRS_{kl} = \dfrac{r}{w} - \dfrac{\lambda}{(1-\lambda)} \dfrac{(s-r)}{w}$	Efficient mix of capital and labor	Overuse capital relative to labor
$MRS_{kx^{k}} = \dfrac{r}{\gamma^{k}} - \dfrac{\lambda}{(1-\lambda)} \dfrac{(s-r)}{\gamma^{k}}$	Efficient mix of capital and capital augmenting R&D	Overuse capital relative to capital augmenting R&D
$MRS_{kx^{k}} = \dfrac{r}{\gamma^{l}} - \dfrac{\lambda}{(1-\lambda)} \dfrac{(s-r)}{\gamma^{l}}$	Efficient mix of capital and labor augmenting R&D	Overuse capital relative to labor augmenting R&D
$MRS_{lx^{k}} = \dfrac{w}{\gamma^{k}}$	Efficient mix of labor and capital augmenting R&D	Efficient mix of labor and capital augmenting R&D
$MRS_{lx^{l}} = \dfrac{w}{\gamma^{l}}$	Efficient mix of labor and labor augmenting R&D	Efficient mix of labor and labor augmenting R&D

and

$$\frac{\partial f}{\partial (l_{i} \chi^{l})} = \frac{\gamma^{l}(1 - \lambda)}{(1 - \lambda)\left(P + f \dfrac{dP}{dq_{i}}\right) l_{i}}.$$

Combining these:

$$
(3.34) \quad \frac{\dfrac{\partial f}{\partial (l_i \chi^l)}}{\dfrac{\partial f}{\partial (k_i \chi^k)}} = \frac{\gamma^l (1-\lambda)}{(1-\lambda)\left(P + f \dfrac{dP}{dq_i}\right) l_i} \cdot \frac{(1-\lambda)\left(P + f \dfrac{dP}{dq_i}\right) k_i}{\gamma^k (1-\lambda)}
$$

$$
= \frac{\gamma^l k_i}{\gamma^k l_i}.
$$

Substituting (*3.34*) into (*3.33*) reveals:

$$
(3.35) \quad \frac{x_i^k}{x_i^l} = \frac{1 - \lambda \dfrac{s}{r}}{(1-\lambda)} \cdot \frac{rk_i}{wl_i} \cdot \frac{\gamma^l}{\gamma^k} - \frac{\sigma \sum x_j^l}{\sigma \sum x_j^k}.
$$

In other words, the ratio of capital augmenting R&D to labor augmenting R&D depends on the degree of regulation (λ), the ratio of capital costs to labor cost, the price of labor augmenting R&D to capital augmenting R&D, as well as the ratio of labor R&D spillovers to capital R&D spillovers.

Proposition 3.8. The regulated firm will overuses labor augmenting R&D relative to capital augmenting R&D.

The result of this proposition is similar to Smith (1974).[5] Using equation (*3.35*), this follows because if the firm is unregulated ($\lambda = 0$), then the ratio of capital augmenting R&D to labor augmenting R&D becomes:

$$
\frac{x_i^k}{x_i^l} = \frac{rk_i}{wl} \cdot \frac{\gamma^l}{\gamma^k} - \frac{\sigma \sum x_j^l}{\sigma \sum x_j^k}.
$$

On the other hand, as regulation becomes more effective:

$$
\frac{x_i^k}{x_i^l} < \frac{rk_i}{wl} \cdot \frac{\gamma^l}{\gamma^k} - \frac{\sigma \sum x_j^l}{\sigma \sum x_j^k}.
$$

because clearly,

$$
\left(1 - \lambda \frac{s}{r}\right) < (1 - \lambda),
$$

$\lambda s < r,$

$0 < \lambda \leq 1,$

and as a result,

$$\frac{1 - \lambda \dfrac{s}{r}}{(1 - \lambda)} < 1.$$

Consequently, in the regulated case, the ratio of capital augmenting R&D to labor augmenting R&D must be less than it was in the unregulated case. Regulation causes the firm to overuse labor augmenting R&D relative to capital augmenting R&D.

Notes

[1] Alternatively, one could define labor more broadly as "noncapital." This alteration has no substantive impact on the model or the results, but does give it greater generality.

[2] Here we are simply assuming that the spillover rate in research is zero. Since this simplifying assumption may not always be true, the case where positive research spillovers are allowed is examined in section 3.2.

[3] The alternative would simply imply that the firm's rate of return is below that of the maximum rate allowed by the regulatory constraint. Hence, in this nonbinding case, the regulatory constraint would simply be of no consequence to the firm and would be ignored. In the maximum conditions that follow, this nonbinding case is equivalent to the case where $\lambda = 0$.

[4] This assumption follows from Smith (1974, 1975) and Okuguchi (1975). See section 2.3.4 from Chapter 2 for a more detailed discussion.

[5] Of course, the method and inclusion of research spillovers is different from Smith. See section 2.3.4 from Chapter 2 for a more detailed discussion of Smith (1974).

Appendix: Note on Neoclassical and Endogenous Technology

Thus far, the specification of production with respect to technology was fully general, yet did assume the firm had endogenous control over the level of technology. That is, equation (*3.2*) stated:

$$q_i = q_i(x_i, k_i, l_i),$$

where q_i is production output, x_i is quantity of technology, k_i is quantity of capital, and l_i is quantity of labor. As it turns out, interpretations of the impact of technology on production are generally divided into two paradigms: neoclassical growth theory, and endogenous growth theory (see for example Romer 1994, Barro and Sala-i-Martin 1995). Whereas neoclassical theory interprets improvements in technology as a consequences of factors outside the firm's control, endogenous growth interprets such improvements as a consequence of factors within the firm's control. For example, the invention of the steam engine in the late 1700s can be seen as a technological improvement that benefited firms in the neoclassical sense, whereas the subsequent refinements of steam engines can be seen as technological improvements in the endogenous sense. The improvements by James Watt and Mathew Boulton on Thomas Newcomen's steam engine constituted a technological improvement which benefited many firms, regardless of weather these firms specifically contributed to the improvements in steam engine technology. In this case the improvement in technology can be considered neoclassical, whereas latter improvements were generally refinements that benefited firms that specifically invested in such technologies.

In this appendix, we investigate the case in which production is neoclassical and neutral with respect to technology, and the case in which production is endogenous and neutral with respect to technology. In both cases, the results show consistency with the previous findings from this chapter.

A.3.1 Neoclassical Production Functions with Technology

The neoclassical production function augmented with an exogenous index of technology can be most simply stated as:

$(A.3.1) \quad q_i = x(t)f(k_i, l_i),$

where t indicates time. Time is important here because it indicates that technology improves for reasons outside the model. Moreover, notice that the level of technology (x) is without subscript i, meaning that technology is fully exogenous to the firm.

Aside from the above, in the class of neoclassical production functions, exogenous technology is generally introduced in one of three ways: as labor-augmenting, capital augmenting, or neutral. As defined by Barro and Sala-i-Martin (1995), these three alternative specifications can be stated as simply:

$(A.3.2) \quad q_i = f(k_i, l_i \cdot x),$

$(A.3.3) \quad q_i = f(k_i \cdot x, l_i),$

$(A.3.4) \quad q_i = x \cdot f(k_i, l_i),$

where $(A.3.2)$ is the labor-augmenting production function, $(A.3.3)$ is the capital-augmenting production function, and $(A.3.4)$ is a Hicks-neutral production function.[1] This Hicks-neutral production function, which is identical to $(A.3.1)$, is certainly more noted than the other two, and is termed neutral because technical change does not impact the marginal substitution of capital and labor.

Consequently, if the impact of technology is neutral on production, the problem for the regulated firm is to maximize profits subject to the rate-of-return regulatory constraint, and the resulting necessary constrained first-order conditions are:

$$0 = (1 - \lambda)\left(P + xf\frac{dP}{dq_i}\right)\frac{\partial f}{\partial k_i}x - r + s\lambda,$$

$$0 = (1 - \lambda)\left[\left(P + xf\frac{dP}{dq_i}\right)\frac{\partial f}{\partial l_i}x - w\right],$$

and

$$0 = P(Q(xf(k_i, l_i)))xf(k_i, l_i)) - sk_i - wl_i.$$

These are combined in a similar manner as before to reveal the marginal substitution of capital for labor:

$$(A.3.5) \quad MRS_{kl} = \frac{MP_k}{MP_l} = \frac{r - s\lambda}{w(1 - \lambda)} = \frac{r}{w} - \frac{\lambda}{(1 - \lambda)}\frac{(s - r)}{w}.$$

The implications from this marginal relationship are similar to those presented earlier. In the neoclassical neutral case, as can be shown in the labor augmenting and capital augmenting cases, the unregulated firm tends to use its capital efficiently relative to labor, while the regulated firm tends to overuse capital relative to labor.

A.3.2 Endogenous Growth Production Functions

In the class of endogenous growth production functions, there are three distinct specifications (see Romer 1994; Barro and Sala-i-Martin 1995). The first variant assumes that investment in labor endogenously increases the level of technology and production. Hence, production for the ith firm takes the form:

$$(A.3.6) \quad q_i = X(L)f(k_i, l_i)$$

where

$$X(L) = x_i(l_i) + \sigma \sum_{i \neq j} x_j(l_j).$$

Labor input from firm i is narrowly defined l_i, and X without subscript denotes the aggregate stock of R&D knowledge which is a function of the aggregate stock of labor L.[2] The spillover rate in research is represented by σ.

The second model assumes that spillovers from capital investments increase the level of technology:

$$(A.3.7) \quad q_i = X(K)f(k_i, l_i)$$

where

$$X(K) = x_i(k_i) + \sigma \sum_{i \neq j} x_j(k_j).$$

As before, x_i represents expenditure on research and development by firm i, and X without subscript denotes the aggregate stock of knowledge, σ represents the spillover rate in research, and K without subscript denotes the aggregate stock of physical capital.[3]

The final variant allows for neutral knowledge spillovers in research and development. Hence, the production function takes the form:

$(A.3.8)$ $q_i = X \cdot f(k_i, l_i)$

where

$$X = x_i + \sigma \sum_{i \neq j} x_j .$$

Again, x_i represents expenditure on research and development by firm i, and X without subscript denotes the aggregate stock of knowledge, and σ represents the spillover rate in research.[4]

These production functions include both an endogenous mechanism for the improvement in technology, as well as a parameter σ which denotes the impact of research and development from other firms in the market. It follows that $0 \leq \sigma \leq 1$, where $\sigma = 1$ indicates the complete absence of research spillovers, and $\sigma = 1$ indicates full research spillover.

Considering just the neutral spillover model, we see again that the problem for the regulated firm is to maximize profits subject to the rate-of-return regulatory constraint, and the resulting necessary constrained first-order conditions are:

$$0 = (1 - \lambda)\left[P(x_i + \sigma)\frac{\partial f}{\partial k_i} + f(x_i + \sigma)^2 \frac{dP}{dq_i} \frac{\partial f}{\partial k_i} \right] - r + s\lambda$$

$$= (1 - \lambda)\left[P + (x_i + \sigma)\frac{dP}{dq_i} f \right](x_i + \sigma)\frac{\partial f}{\partial k_i} - r + s\lambda ,$$

$$0 = (1 - \lambda)\left[P(x_i + \sigma)\frac{\partial f}{\partial l_i} + f(x_i + \sigma)^2 \frac{\partial P}{\partial q_i} \frac{\partial f}{\partial l_i} - w \right]$$

$$= (1 - \lambda)\left[P + (x_i + \sigma)\frac{\partial P}{\partial q_i} f \right](x_i + \sigma)\frac{\partial f}{\partial l_i} - w ,$$

$$0 = (1 - \lambda)\left[Pf + f^2(x_i + \sigma)\frac{\partial P}{\partial q_i} - \gamma \right]$$

$$= (1 - \lambda)\left[P + (x_i + \sigma)\frac{\partial P}{\partial q_i} f \right]f - \gamma$$

and

$$0 = P(Q(X \cdot f(k_i, l_i)))X \cdot (k_i, l_i) - sk_i - wl_i - \gamma x_i .$$

These first order conditions can be rearranged as marginal products and then combined to reveal the marginal rates of substitution:

$$(A.3.9) \quad MRS_{kl} = \frac{MP_k}{MP_l} = \frac{r - s\lambda}{(1-\lambda)\left[P + (\sigma + x_i)\dfrac{\partial P}{\partial q_i} f\right]}$$

$$\cdot \frac{(1-\lambda)\left[P + (\sigma + x_i)\dfrac{\partial P}{\partial q_i} f\right]}{(1-\lambda)w}$$

$$= \frac{r - s\lambda}{w(1-\lambda)} = \frac{r}{w} - \frac{\lambda}{(1-\lambda)}\frac{(s-r)}{w},$$

$$(A.3.10) \quad MRS_{kx} = \frac{MP_k}{MP_x} = \frac{r - s\lambda}{(1-\lambda)\left[P + (\sigma + x_i)\dfrac{\partial P}{\partial q_i} f\right]}$$

$$\cdot \frac{(1-\lambda)\left[P + (\sigma + x_i)\dfrac{\partial P}{\partial q_i} f\right]}{(1-\lambda)\gamma}$$

$$= \frac{r - s\lambda}{\gamma(1-\lambda)} = \frac{r}{\lambda} - \frac{\lambda}{(1-\lambda)}\frac{(s-r)}{\gamma},$$

and

$$(A.3.11) \quad MRS_{lx} = \frac{MP_l}{MP_x} = \frac{(1-\lambda)w}{(1-\lambda)\left[P + (\sigma + x_i)\dfrac{\partial P}{\partial q_i} f\right]}$$

$$\cdot \frac{(1-\lambda)\left[P + (\sigma + x_i)\dfrac{\partial P}{\partial q_i} f\right]}{(1-\lambda)\gamma}$$

$$= \frac{w(1-\lambda)}{\gamma(1-\lambda)} = \frac{w}{\gamma}.$$

The implications from these marginal relationships are identical to the augmented Averch-Johnson case before. The regulated firm will use a larger ratio of capital to labor and capital to R&D than the unregulated firm. In terms of labor and R&D, however, both regulated and unregulated firms act similarly.

Notes

[1] Time is ignored at this point.

[2] The origin of this labor-spillover production function is generally attributed to Robert Lucas (1988). Lucas in fact defines l as the aggregate stock of human capital and l_i as simply human capital input.

[3] This capital-spillover model is generally attributed to Kenneth Arrow (1962).

[4] This knowledge-spillover model is generally attributed to Paul Romer (1986, 1994).

4 Research Joint Ventures

In this chapter, a symmetric two-stage Nash equilibrium duopoly model of cost reducing R&D is considered in which each of the firms is subject to rate-of-return regulation. In the initial stage firms can choose either to cooperate or not to cooperate in the R&D market, while in the final stage firms can choose either to cooperate or not to cooperate in the output market. Three specific cases are examined: (1) the case in which firms compete against each other in both the research and output stages (2) the case in which firms form a research joint venture in the first stage but remain noncooperative in the final stage, and (3) the case in which firms cooperate in both R&D and output.

The models used here closely follow the models of cooperative and noncooperative R&D spillovers put forth by Claude D'Aspremont and Alexis Jacquemin (1988), Raymond De Bondt, Patrick Slaets, and Bruno Cassiman (1992), Kotaro Suzumura (1992), and Sang-Seung Yi (1996). In all these instances, a two-stage game with R&D spillovers is used consisting of R&D cooperation or competition in the initial stage, followed by output cooperation or competition in the final stage. Michael Katz (1986) also provides an excellent but broader four-stage model of research cooperation and production cooperation. Joanna Poyago-Theotoky (1995) investigates a two-stage research joint venture model but is narrowly concerned with the equilibrium size of the joint venture, which is found to be less than optimal at equilibrium.

D'Aspremont and Jacquemin differ from the rest by specifically investigating the symmetric *duopoly* case with homogenous goods. They conclude that cooperation in R&D generally *increases* the firm's level of R&D expenditures and output expenditures.[1] They note that this is somewhat remarkable considering the general expectation that joint ventures in research should *decrease* R&D expenditures because of less wasteful duplications. This outcome, however, likely reflects the appropriability of R&D benefits to the individual firm. When spillovers are sufficiently high, competing firms have little incentive to enact costly

78

R&D investments since the benefits from doing so will not be exclusive to themselves. Once engaged in a R&D joint venture, however, this incentive changes and R&D expenditures increase.

Suzumura (1992) deals with the more general two-stage case in which the industry consists of two *or more* firms producing a homogenous product. Suzumura then compares these results with the first-best and second-best welfare benchmarks. The basic finding is that R&D cooperation tends to increase R&D spending. Although this conclusion is not substantially different from the previous findings, Suzumura does find that if research spillovers between firms are sufficiently large, both full noncooperative and R&D cooperative outcomes are socially inefficient from a first-best and second-best welfare standard. However, in the absence of spillovers, Suzumura finds that the outcome from R&D cooperation remains socially inefficient, while the level of R&D from noncooperative equilibrium is beyond the socially efficient level. Suzumura argues that this occurs because without cooperation, the appropriability of R&D benefits is small, and thus when spillovers are sufficiently large, R&D cooperation tends to increase the level of R&D expenditures. Only when there are no research spillovers do competing firms produce a more than socially efficient level of R&D.

De Bondt, Slaets, and Cassiman (1992) use the same basic symmetric two-stage Nash equilibrium model as Suzumura and D'Aspremont and Jacquemin, but focus extensively on the effects of spillovers, the number of rivals, and product differentiation on R&D and production output. Although their conclusions are in many respects similar to Suzumura and D'Aspremont and Jacquemin, they find that a large number of rivals reduces R&D investments so long as research spillovers are sufficiently small and products are homogenous. When spillovers are moderate, effective R&D increases with the number of rivals. However, both of these trends may be reversed if goods are sufficiently differentiated.

Sang-Seung Yi (1996) provides a useful two-stage model of cooperative and noncooperative R&D which incorporates an arbitrary number of firms, variations in the elasticity of demand, various spillover rates, and differs from the above by focusing mainly on the social and consumer welfare implications. Specifically, Yi finds that cooperation reduces R&D when spillovers are low or intermediate, and reduces social welfare when spillovers are intermediate (but not low). However, as the elasticity of demand increases, Yi finds that cooperation tends to *increase* R&D and social welfare, and in the limit, cooperation is beneficial for all spillover rates. As before, this outcome likely reflects the appropriability of R&D benefits to the individual firm. When the elasticity of demand is sufficiently low and R&D spillovers are sufficiently high, competing firms have little incentive to enact costly R&D investments since the benefits from doing so will not be exclusive to themselves. However, once engaged in a R&D joint venture, R&D expenditures increase because this incentive changes. As the elasticity of demand increases (approaches zero), the firm has less market power and consequently the appropriability problem exists for a even lower level of R&D spillovers. Hence, in the limit, when the elasticity of demand is zero, R&D cooperation is socially beneficial for all spillover levels.

The critical difference between these models and the model here is the presence of rate-of-return regulation. We consider an industry with two firms producing a homogenous product with a linear inverse demand curve:

$(4.1) \qquad P = a - bQ,$

where P is the market price and Q is the total industry output, such that

$$Q = q_1 + q_2.$$

It is assumed that $a, b > 0$ and $a / b \geq Q$. Following D'Aspremont and Jacquemin (1988), De Bondt, Slaets, and Cassiman (1992), and Poyago-Theotoky (1995), it is assumed that the cost of firm i's own R&D reflects the existence of diminishing returns to R&D expenditures:

$(4.2) \qquad \gamma \dfrac{x_i^2}{2}$

where γ is the price of R&D. For simplicity, it is also assumed that every unit of expenditure on R&D generates one unit of effective cost-reducing R&D.

We assume that per unit production cost in the absence of R&D is $rk_i + wl_i$ where k_i and l_i are the capital and labor required to produce one unit of output. The production cost, however, may be lowered through cost reducing R&D so that, taking R&D into consideration, we may specify the per unit production as

(4.3) $c_i = rk_i + wl_i - x_i - \sigma x_j$,

where x_i is cost reducing R&D from the ith firm and x_j is cost reducing R&D from the jth firm. Notice that σ is the exogenous R&D spillover parameter, if $\sigma = 0$ then there are no R&D spillovers, if $\sigma = 1$, then there are perfect R&D spillovers. It is assumed that $rk_i + wl_i > x_i + \sigma x_j$ because cost cannot be negative.

Using this framework, the following sections will consider the impact of rate-of-return regulation on R&D for each of the three models of research cooperation. Section 4.1 considers the case in which firms compete against each other in both the research and output stages, section 4.2 considers the case in which firms form a research joint venture in the first stage but remain noncooperative in the final stage, and section 4.3 considers the case in which firms cooperate in both R&D and output. The concluding section brings the results from all three of these cases together and comparatively evaluates each with respect to R&D expenditure and production output.

4.1 Noncooperative R&D and Output

If firms compete in both stages of the game, then profit for the ith firm in the second stage is

$$\Pi_i = (a - bQ)q_i - \left(rk_i + wl_i - x_i - \sigma x_j\right)q_i - \gamma \frac{x_i^2}{2},$$

where $j \neq i$ and $i = 1, 2$. The imposition of rate-of-return regulation means that

$$(a - bQ)q_i - \left(sk_i + wl_i - x_i - \sigma x_j\right)q_i - \gamma \frac{x_i^2}{2} = 0,$$

where s is the allowed cost of capital as defined in Chapter 2 and 3. Consequently, the problem for the regulated firm is to maximize profits subject to the regulatory constraint:

(4.4) $L = (a - bQ)q_i - \left(rk_i + wl_i - x_i - \sigma x_j\right)q_i - \gamma \frac{x_i^2}{2}$

$$-\lambda\left[(a-bQ)q_i-\left(sk_i+wl_i-x_i-\sigma x_j\right)q_i-\gamma\frac{x_i^2}{2}\right],$$

where $0 \leq \lambda \leq 1$, and $s > r$. Recall from Chapter 3 that λ reflects the degree and effectiveness of regulation. With perfect rate-of-return regulation, λ will equal one, but with perfectly ineffective regulation, λ will equal zero. Moreover, as λ approaches one, regulation becomes more effective, and as λ falls towards zero, regulation becomes less effective. The assumption that $s > r$ follows from the original Averch-Johnson model, and was used in Chapter 3. It is also common to extend this assumption to $r > s\lambda$, as was done in Chapter 3 section 2, and in Smith (1974, 1985) and Okuguchi (1975).[2]

Since

$$\frac{\partial L}{\partial q_i}=(1-\lambda)(a-bq_i-bQ-wl_i+x_i+\sigma x_j)-k(r-s\lambda)$$

$$= 0,$$

then the Nash-Cournot symmetric equilibrium for this second stage is

$$(4.5) \qquad q_i^{NN}=\frac{1}{3b}\left[a-wl_i+x_i+\sigma x_j-\frac{k_i(r-s\lambda)}{1-\lambda}\right].$$

If q_i^{NN} were negative, the firm would not produce. Therefore we assume that q_i^{NN} is positive. This in turn requires

$$a-wl_i-\frac{k_i(r-s\lambda)}{1-\lambda}$$

is positive.

The first stage payoff function for firm i is defined as profits (4.4) at the second stage equilibrium output (4.5):

$$\Pi_i^{NN}=\frac{1}{9b}\left\{\left(a^2+2w^2l_i^2+2wl_ix_i-4.5bx_i^2\gamma-2wl_i\sigma x_j+\sigma^2x_j^2\right)\right.$$

$$\cdot(1-\lambda)-2a[(wl_i-x_i-\sigma x_j)(1-\lambda)+k_i(r-s\lambda)]$$

$$\left.+2k_i(wl_i-x_i-\sigma x_j)(r-s\lambda)+\frac{k_i^2(r-s\lambda)^2}{1-\lambda}\right\}.$$

Since

$$\frac{\partial \Pi_i^{NN}}{\partial x_i} = \frac{\left(a - wl_i - x_i + 4.5bx_i\gamma - \sigma x_j\right)\left(1 - \lambda\right) - k_i\left(r - s\lambda\right)}{4.5b},$$

then the symmetric solution for the ith firm's R&D is

$$(4.6) \qquad x_i^{NN} = \frac{1}{4.5b\gamma - (1+\sigma)}\left[a - wl_i - \frac{k_i\left(r - s\lambda\right)}{1 - \lambda}\right].$$

After substituting (4.6) into (4.5), we can solve for the symmetric market output solution:

$$(4.7) \qquad Q^{NN} = q_i^{NN} + q_j^{NN} = \frac{3\gamma}{4.5b\gamma - (1+\sigma)}\left[a - wl_i - \frac{k_i\left(r - s\lambda\right)}{1 - \lambda}\right].$$

Notice from equation (4.6) the effect of regulation (λ) on R&D (x_i^{NN}): as regulation increases ($\lambda \to 1$) R&D falls, but as regulation decreases ($\lambda \to 0$) R&D increases. This occurs if $4.5b\gamma > \sigma + 1$, and because it is assumed that the term in brackets is positive (so that q_i^{NN} will be positive), it must be true that $4.5b\gamma > \sigma + 1$ in order for Q^{NN} to be positive. Consequently,

$$\underset{\lambda \to 1}{Lim}\left\{\frac{1}{4.5b\gamma - (1+\sigma)}\left[a - wl_i - \frac{k_i\left(r - s\lambda\right)}{1 - \lambda}\right]\right\} = -\infty,$$

precisely because

$$\underset{\lambda \to 1}{Lim}\left[-\frac{k_i\left(r - s\lambda\right)}{1 - \lambda}\right] = -\infty.$$

On the other hand, the limit as λ approaches zero:

$$\underset{\lambda \to 0^+}{Lim}\left\{\frac{1}{4.5b\gamma - (1+\sigma)}\left[a - wl_i - \frac{k_i\left(r - s\lambda\right)}{1 - \lambda}\right]\right\}$$

$$= \frac{1}{4.5b\gamma - (1+\sigma)}\left[a - wl_i - rk_i\right] > 0.$$

Hence, as rate-of-return regulation increases (decreases), *ceteris paribus*, expenditures on R&D decrease (increase). This result is similar to the Averch-Johnson effect explored in Chapters 2 and 3. With technology as an addition factor of production, we not only see that rate-of-return regulation causes an overuse of capital relative to labor, but also an overuse of capital relative to technology. Hence, as regulation increases, expenditures on R&D decrease. What remains is to compare this effect to the case where the firms form a joint research venture in the first stage but remain noncooperative in the output stage, and the case where the firms cooperate in both R&D and output.

4.2 Cooperative R&D and Noncooperative Output

If firms cooperate in the R&D market but remain competitive in the output market, then, as before, profit for the ith firm in the second stage is

$$\Pi_i = (a - bQ)q_i - (rk_i + wl_i - x_i - \sigma x_j)q_i - \gamma \frac{x_i^2}{2},$$

where $j \neq i$ and $i = 1, 2$. The imposition of rate-of-return regulation means that

$$(a - bQ)q_i - (sk_i + wl_i - x_i - \sigma x_j)q_i - \gamma \frac{x_i^2}{2} = 0.$$

Consequently, the problem for the regulated firm is to maximize profits subject to the regulatory constraint:

$$(4.8)\ L = (a - bQ)q_i - (rk_i + wl_i - x_i - \sigma x_j)q_i - \gamma \frac{x_i^2}{2}$$

$$- \lambda \left[(a - bQ)q_i - (sk_i + wl_i - x_i - \sigma x_j)q_i - \gamma \frac{x_i^2}{2} \right],$$

where $0 \leq \lambda \leq 1$, and $s > r$. Since

$$\frac{\partial L}{\partial q_i} = (1 - \lambda)(a - bq_i - bQ - wl_i + x_i + \sigma x_j) - k(r - s\lambda) = 0,$$

then the Nash-Cournot symmetric equilibrium for this second stage is

$$(4.9)\ q_i^{CN} = \frac{1}{3b} \left[a - wl_i + x_i + \sigma x_j - \frac{k_i(r - s\lambda)}{1 - \lambda} \right].$$

The first stage payoff function for firm i is defined as the sum of firm i's and firm j's profit (4.8) at the second stage equilibrium output (4.9):

$$\Pi^{CN} = \Pi_i^{CN} + \Pi_j^{CN} = \frac{1}{4.5b}\Big\{a(1-\lambda) - w^2 l_i^2 - x_i(1 - 4.5b\gamma)$$
$$- 2x_i\sigma x_j - \sigma^2 x_j^2 + 2wl_i(x_i + \sigma x_j) + 2k_i(wl_i - x_i - \sigma x_j)$$
$$\cdot(r - s\lambda) - 2a[(wl_i - x_i - \sigma x_j)(1-\lambda) + k_i(r - s\lambda)]$$
$$+ \frac{k_i^2(r - s\lambda)^2}{1-\lambda}\Big\}.$$

Since

$$\frac{\partial \Pi^{CN}}{\partial x_i} = \frac{1}{4.5b}\Big\{\big(-2wl_i + 2x_i - 9bx_i\gamma - 2wl_i\sigma + 4x_i\sigma - 2x_i\sigma^2\big)$$
$$\cdot(1-\lambda) + 2(1+\sigma)[a(1-\lambda) - k_i(r - s\lambda)]\Big\},$$

then optimal R&D for the ith firm is

(4.10) $\quad x_i^{CN} = \frac{(1+\sigma)}{4.5b\gamma - (1+\sigma)^2}\left[a - wl_i - \frac{k_i(r - s\lambda)}{1-\lambda}\right].$

After substituting (4.10) into (4.9), we can solve for the symmetric market output solution:

(4.11)

$$Q^{CN} = q_i^{CN} + q_j^{CN} = \frac{3\gamma}{4.5b\gamma - (1+\sigma)^2}\left[a - wl_i - \frac{k_i(r - s\lambda)}{1-\lambda}\right].$$

Notice from (4.10) the effect of regulation (λ) on R&D (x_i^{CN}): as regulation increases ($\lambda \to 1$) R&D falls, but as regulation decreases ($\lambda \to 0$) R&D increases. Consequently,

$$\underset{\lambda \to 1}{Lim}\left\{\frac{1}{4.5b\gamma - (1+\sigma)}\left[a - wl_i - \frac{k_i(r - s\lambda)}{1-\lambda}\right]\right\} = -\infty,$$

precisely because

$$\underset{\lambda \to 1}{Lim}\left[-\frac{k_i(r - s\lambda)}{1-\lambda}\right] = -\infty.$$

On the other hand, the limit as λ approaches zero:

$$Lim_{\lambda \to 0^+} \left\{ \frac{1}{4.5b\gamma - (1+\sigma)} \left[a - wl_i - \frac{k_i(r - s\lambda)}{1 - \lambda} \right] \right\}$$

$$= \frac{1}{4.5b\gamma - (1+\sigma)} [a - wl_i - rk_i] > 0 .$$

Consequently, as rate-of-return regulation increases (decreases), *ceteris paribus*, expenditures on R&D decrease (increase). This conclusion is similar to the previous conclusion from the full noncooperative case. Essentially, the presence of rate-of-return regulation causes an underuse of R&D, although the magnitude of this effect on R&D is different from the full noncooperative case of section 4.1.

4.3 Cooperative R&D and Output

If firms cooperate in both stages of the game, then profit for the firms in the second stage is

$$\Pi_i = (a - bQ)Q - (rk_i + wl_i)Q + (x_i + \sigma x_j)q_i + (x_j + \sigma x_i)q_j$$

$$- \gamma \frac{x_i^2}{2} - \gamma \frac{x_j^2}{2} ,$$

where $j \neq i$ and $i = 1, 2$. The imposition of rate-of-return regulation means that

$$(a - bQ)Q - (sk_i + wl_i)Q + (x_i + \sigma x_j)q_i + (x_j + \sigma x_i)q_j$$

$$- \gamma \frac{x_i^2}{2} - \gamma \frac{x_j^2}{2} = 0 ,$$

where s is the allowed cost of capital. Therefore, the problem for the regulated firm is to maximize profits subject to the regulatory constraint:

$$(4.12) \quad L = (a - bQ)Q - (rk_i + wl_i)Q + (x_i + \sigma x_j)q_i + (x_j + \sigma x_i)q_j$$

$$- \gamma \frac{x_i^2}{2} - \gamma \frac{x_j^2}{2} - \lambda [(a - bQ)Q - (sk_i + wl_i)Q$$

$$+ (x_i + \sigma x_j)q_i + (x_j + \sigma x_i)q_j - \gamma \frac{x_i^2}{2} - \gamma \frac{x_j^2}{2}] ,$$

where $Q = q_i + q_j$, $0 \leq \lambda \leq 1$, and $s > r$. In the symmetric case,

$$\frac{\partial L}{\partial Q} = (1 - \lambda)\left[a - 4bQ - wl_i + \frac{x_i}{2}(1 + \sigma) + \frac{x_j}{2}(1 + \sigma)\right]$$
$$- k_i(r - s\lambda) = 0,$$

and thus the Nash-Cournot equilibrium for the second stage is

(4.13) $\quad Q^{CC} = \dfrac{1}{2b}\left[a - wl_i + \dfrac{x_i}{2}(1 + \sigma) + \dfrac{x_j}{2}(1 + \sigma) - \dfrac{k_i(r - s\lambda)}{1 - \lambda}\right]$.

The first stage payoff function is defined as profits (4.12) at the second stage equilibrium output (4.13)

$$\Pi^{CC} = -\frac{1}{4b}\left\{\left[a^2 + w^2 l_i^2 - 2wl_i x(1 + \sigma) - 4x^2 b\gamma + x^2(1 + \sigma)^2\right](1 - \lambda)\right.$$

$$\left[2k_i(r - s\lambda) + 2a(1 - \lambda)\right]\left[wl_i - x(1 + \sigma)\right] - 4ak_i(r - s\lambda)$$
$$\left. + \frac{k_i^2(r - s\lambda)^2}{1 - \lambda}\right\}.$$

Since

$$\frac{\partial \Pi^{CC}}{\partial x} = \frac{1}{2b}\left\{\left[a(1 + \sigma) - wl_i(1 + \sigma) + x(1 - 4b\gamma + 2\sigma - \sigma^2)\right]\right.$$
$$\left. \cdot (1 - \lambda) - k_i(r - s\lambda)(1 + \sigma)\right\},$$

then the symmetric solution $(x_i = x_j = x)$ for the firm's R&D is

(4.14) $\quad x^{CC} = \dfrac{(1 + \sigma)}{4b\gamma - (1 + \sigma)^2}\left[a - wl_i - \dfrac{k_i(r - s\lambda)}{1 - \lambda}\right]$.

After substituting (4.14) into (4.13), we can solve for the symmetric market output solution:

(4.15) $\quad Q^{CC} = \dfrac{2\gamma}{4b\gamma - (1 + \sigma)^2}\left[a - wl_i - \dfrac{k_i(r - s\lambda)}{1 - \lambda}\right]$.

As in the previous two cases, regulation (λ) inversely affects R&D (x^{CC}). This follows from equation (*4.14*) because we assume $r > s\lambda$. Consequently,

$$Lim_{\lambda \to 1}\left\{\frac{1}{4b\gamma - (1+\sigma)^2}\left[a - wl_i - \frac{k_i(r-s\lambda)}{1-\lambda}\right]\right\} = -\infty,$$

precisely because

$$Lim_{\lambda \to 1}\left[-\frac{k_i(r-s\lambda)}{1-\lambda}\right] = -\infty.$$

However, as λ approaches zero:

$$Lim_{\lambda \to 0^+}\left\{\frac{1}{4b\gamma - (1+\sigma)^2}\left[a - wl_i - \frac{k_i(r-s\lambda)}{1-\lambda}\right]\right\}$$

$$= \frac{1}{4b\gamma - (1+\sigma)^2}[a - wl_i - rk_i] > 0.$$

Therefore, as rate-of-return regulation increases (decreases), *ceteris paribus*, expenditures on R&D decrease (increase). The presence of rate-of-return regulation causes an underuse of technology. This is, of course, the general result in each of the three cooperative models. What varies among the cases is the magnitude of this effect on R&D expenditures. The following section will evaluate this issue by comparing the three cases.

4.4 Discussion

By extending the research joint venture model to the case in which firms are subject to rate-of-return regulation, we conclude that although cooperation in production and research tends to increase R&D, the presence of rate-of-return regulation tends to decrease both production output and R&D. Looking at R&D in isolation, this follows from equations (*4.6*), (*4.10*), and (*4.14*). As long as some degree of research spillovers are present ($\sigma > 0$), then it must be the case that:

$$x_i^{CC} > x_i^{CN} > x_i^{NN},$$

precisely because

$$\frac{(1+\sigma)}{4b\gamma - (1+\sigma)^2} > \frac{(1+\sigma)}{4.5b\gamma - (1+\sigma)^2} > \frac{1}{4.5b\gamma - (1+\sigma)},$$

as long as

$$b \geq \frac{1}{\gamma}.$$

Recall that parameter b is the slope from the linear inverse demand function (4.1) and is by definition always greater than zero. The parameter γ is the price of R&D to the firm and, like b, is by definition always greater than zero. The general implication from this is that $x_i^{CC} > x_i^{CN} > x_i^{NN}$, as long as b is not too close to zero. If, however, $b < 1 / \gamma$, then for the very high spillover levels ($\sigma > 0.75$) it may be the case that $x_i^{CN} < x_i^{NN}$. Of course, if there are no research spillovers ($\sigma = 0$), then for all cases, $x_i^{CC} > x_i^{CN} = x_i^{NN}$.

The intuition behind this result is that the impact of regulation on R&D expenditures across the three cooperative models depends on the slope to the inverse demand curve (b), the price of R&D (γ), and the degree of research spillovers (σ). Although from a different context, this interrelationship between demand, research spillovers, and R&D expenditures, is in many respects, similar to the findings of Yi (1996). As long as the slope of demand (b) is sufficiently large and there are some spillovers, R&D from the full cooperative environment will be greater than R&D from the research cooperation and production competition environment, which in turn will be greater than R&D from the full noncooperative environment. Figure 4.1 provides a graphical example of this relationship between the degree of research spillovers (σ) and the various levels of R&D.

Looking at just production, it is clear that research joint ventures are beneficial as long as they are not accompanied by output cooperation. This follows from equations (4.7) (4.11), and (4.15). As long as

$$b \geq \frac{2.67}{\gamma},$$

then

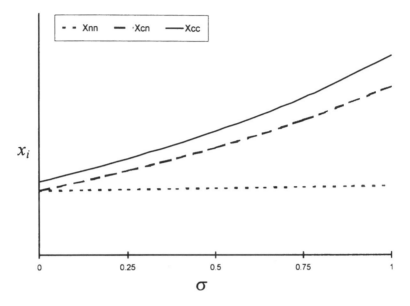

Figure 4.1 The Effect of Research Spillovers on R&D

$$Q^{CN} > Q^{NN} > Q^{CC},$$

precisely because

$$\frac{3\gamma}{4.5b\gamma - (1+\sigma)^2} > \frac{3\gamma}{4.5b\gamma - (1+\sigma)} > \frac{2\gamma}{4b\gamma - (1+\sigma)^2}$$

for all cases where $\sigma > 0$. Hence, with sufficiently high demand slope (b), and any nonzero amount of research spillovers (σ), production output from research cooperation and production noncooperation will be greater than output from the full noncooperative environment, which in turn will be greater than output from the full cooperative environment. Figure 4.2 provides a graphical example of this case. However, if $b\gamma < 2.67$ then for the very high spillover levels ($\sigma > 0.75$), $Q^{NN} < Q^{CC}$. Of course, in all cases, if $\sigma = 0$, then $Q^{CN} = Q^{NN} > Q^{CC}$. Thus, the impact of rate-of-return regulation on production output across the three cooperative cases depends on the slope to the inverse demand curve (b),

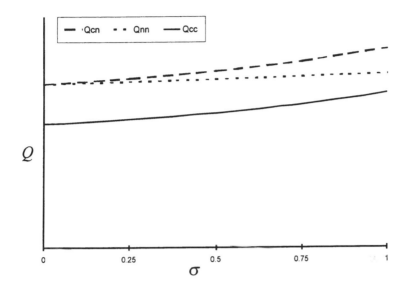

Figure 4.2 The Effect of Research Spillovers on Output

the price of R&D (γ), and the degree of research spillovers (σ). As long as the slope of demand is sufficiently high and there are nonzero spillovers in research, output is greatest when firms are cooperative in research but uncooperative in production.

The key implications are that (1) rate-of-return regulation has the effect of reducing both R&D and production output in all three cases (2) the various forms of cooperation have the effect of increasing R&D, and (3) output is greatest when firms are cooperative in research but noncooperative production. Although regulation reduces R&D, this reduction can be mitigated by participation in joint research ventures with rival firms. These results may or may not be robust to variations in the elasticity of demand or in the number of firms in the market place, however (see Yi 1996, and De Bondt, Slaets, and Cassiman 1992). These variations are important areas for further research.

Notes

[1] They also note that full cooperation in both research and production underperforms both the research cooperative and full noncooperative cases in terms of output, and outperforms both in terms of R&D expenditures.

[2] See section 2.3.4 from Chapter 2 for a more detailed discussion of this assumption.

5 Evidence From Electric Utilities in Texas

We now use a sample of annual observations from ten privately owned electric utilities in the state of Texas covering the years 1965 to 1985 to construct an empirical test the impact of rate-of-return regulation on technical change. Texas is an interesting case because of the large size and capacity of its electric utilities, and because statewide regulation did not begin until 1975 when Texas lawmakers passed the Public Utility Regulatory Act (PURA). Before this period, local governments and municipalities retained regulatory jurisdiction, although most utility companies faced little, if any, regulatory oversight. Beginning in 1937, appeals for electric, telephone, and water rates for cities of all sizes were made directly to the courts. This piecemeal approach lasted until the PURA became effective on September 1, 1975. In describing the period between the 1930s and 1975, Jack Hopper notes that "many utility services remained unregulated: cities had no power to control rates in rural areas; industrial electric and gas rates within cities became private contractual matters in the absence of municipal regulation" (Hopper 1976, 779). In the words of the original legislation:

> This Act is enacted to protect the public interest inherent in rates and services of public utilities. The legislation finds that public utilities are by definition monopolies in the areas they serve; that therefore the normal forces of competition which operate to regulate prices in a free enterprise society do not operate; and that therefore utility rates, operations and services are regulated by public agencies with the objective that such regulation shall operate as a substitute for such competition (*Public Utility Regulatory Act* 1975, 2327-2328).

Dan Pleitz and Robert Little argue that before this act, even regulatory-minded municipalities simply "could not effectively regulate one part of a massive network and facilities stretching throughout the state" (Pleitz and Little 1976, 977). Supporters of the legislation were further dissatisfied by the poor service provided to rural customers. As Jack Hopper and Eric Hartman note, supporters had "reached the point where they were willing to pay almost any price if it would get them reliable service" (Hopper and Hartman 1979, 7). At the same time, public demand for regulation was additionally heightened by the dramatic increases in utility rates resulting from the energy crisis and high inflation (see Gayle 1976).[1] Opposition to this new regulation, although ultimately unsuccessful, was led by the Texas Municipal League, many local and area chambers of commerce, and much of the newspaper, radio, and television medium (see Adams 1976).

As a result, statewide use of rate-of-return regulation in Texas began in 1975. Currently, the Public Utility Commission of Texas retains statewide regulatory authority, but with the signing of Senate Bill 7 in June of 1999 by the governor, this is set to change dramatically. Under the current plan, retail competition will begin for most investor-owned utilities in January of 2002.

Previous empirical studies involving rate-of-return regulation have focused primarily on the Averch-Johnson effect; only a select few have attempted to test the impact of regulation on technical change. Among these are Randy Nelson (1984) and Gerald Granderson (1999). Nelson used a sample of forty firms in the electric power industry who were engaged in the interstate sale of electricity over the years 1951-1978. During this time period, all of these firms were subject to rate-of-return regulation by the Federal Power Commission (FPC). After estimating a translog cost function and share equations, Nelson found that over this period, regulation had a very small but negative effect on the rate of technical change.

Similarly, Gerald Granderson (1999) looked at the impact of rate-of-return regulation on technical change for twenty U.S. interstate natural gas pipeline companies over the years 1977-1987. He utilized estimates from a translog cost function to test the theoretical propositions put forth in Smith (1974, 1975) concerning the impact of rate-of-return regulation on the selection of factor augmenting technology, and concluded that regulation caused firms to adopt technology that augmented noncapital more than capital, and led to a small decrease in the rate of technical change.[2]

Aside from these two articles, all other known empirical attempts involving rate-of-return regulation ignore the question of its impact on technical change. Mathios and Rogers (1989) and Kaestner and Kahn (1990) are two examples. Hence, the contribution of this chapter is to expand on the empirical efforts put forth by Nelson and Granderson, and to provide empirical evidence concerning the impact of rate-of-return regulation on the innovativeness of electric utilities in Texas.

To this end, the following section (section 5.1) defines the translog cost function and resulting share equations which are to be estimated. Section 5.2 defines the construction of the variables, and section 5.3 presents the results from the empirical estimation. Finally, section 5.4 provides a brief summary and conclusion.

5.1 Translog Cost Model

Following Christensen, Jorgenson, and Lau (1971), Nelson (1984), Berndt (1991), and Granderson (1999), a cost function of arbitrary form can be approximated by using a second-order Taylor's series expansion in logarithms.[3] Accordingly, let the unregulated cost function be:

$$C^u = f(Q, P, x)$$

where C^u is total cost, Q is the level of output, P is a vector of the factor input prices, and x is the firm's technology. In the electric utility industry, we assume there are three relevant input factors: labor (l), capital (k), and fuel (f) (see for example Nelson 1984, Kwan 1987, Kumbhakar 1996). This unregulated cost function can then be approximated by the transcendental logarithmic cost function:

$$(5.1) \qquad \ln C^u = \alpha_o + \sum_i \alpha_i \ln P_i + \frac{1}{2} \sum_i \sum_j \alpha_{ij} \ln P_i \ln P_j$$

$$+ \beta_Q \ln Q + \frac{1}{2} \beta_{QQ} (\ln Q)^2 + \sum_i \beta_{Qi} \ln Q \ln P_i$$

$$+ \delta_x x + \delta_{Qx} (\ln Q) x + \frac{1}{2} \delta_{xx} x^2 + \sum_i \delta_{xi} (\ln P_i) x,$$

where $i, j = k, l, f,$ and ln is the natural log operator.

The optimal cost-minimizing share equations can be found by employing Shephard's Lemma and differentiating (*5.1*) with respect to the natural log of the input prices. This follows because

$$d \ln C^u = \frac{1}{C^u} dC^u,$$

$$d \ln P_i = \frac{1}{P_i} dP_i,$$

and as a result,

$$(5.2) \qquad \frac{\partial \ln C^u}{\partial \ln P_i} = \frac{\partial C^u}{\partial P_i} \frac{P_i}{C^u}.$$

Shephard's Lemma states that the optimal share for each input $\left(q_i\right)$ can be found by either the derivative of cost with respect to the price of the input $\left(\partial C^u / \partial P_i\right)$, or by the derivative of the profit maximizing Lagrangian equation with respect to the price of the input $\left(\partial L^u / \partial P_i\right)$. Using the former,

$$\frac{\partial C^u}{\partial P_i} = q_i,$$

along with equation (*5.2*),

$$\frac{\partial \ln C^u}{\partial \ln P_i} = \frac{\partial C^u}{\partial P_i} \frac{P_i}{C^u},$$

we find the cost share for input i,

$$\frac{\partial \ln C^u}{\partial \ln P_i} = \frac{P_i q_i}{C^u} = S_i^u .^4$$

Hence, finding the derivative of (*5.1*) with respect to lnP_i and employing Shepherd's Lemma yields the cost-minimizing share equations,

$$(5.3) \qquad S_k^u = \left(C_k^u\right)' = \frac{\partial \ln C^u}{\partial \ln P_k}$$

$$= \alpha_k + \alpha_{kk} \ln P_k + \alpha_{kl} \ln P_l + \alpha_{kf} \ln P_f,$$

$$+ \beta_{Qk} \ln Q + \delta_{xk} x,$$

$$(5.4) \quad S_l^u = \left(C_l^u\right)' = \frac{\partial \ln C^u}{\partial \ln P_l}$$

$$= \alpha_l + \alpha_{lk} \ln P_k + \alpha_{ll} \ln P_l + \alpha_{lf} \ln P_f$$

$$+ \beta_{Ql} \ln Q + \delta_{xl} x,$$

and

$$(5.5) \quad S_f^u = \left(C_f^u\right)' = \frac{\partial \ln C^u}{\partial \ln P_f}$$

$$= \alpha_f + \alpha_{fk} \ln P_k + \alpha_{fl} \ln P_l + \alpha_{ff} \ln P_f$$

$$+ \beta_{Qf} \ln Q + \delta_{xf} x.$$

Therefore, in the unregulated case, each cost share equation (S_i^u) is given in by the equations (5.3), (5.4), and (5.5). Along with the cost equation (5.1), these three cost share equations constitute a system of equations that can then be estimated.

The above is simply the unregulated case. If rate of return regulation is imposed upon the firm, then the translog cost function and the resulting cost share equations must be altered. Accordingly, let the regulated cost function be:

$$C^r = f(Q, P, s, x)$$

where C^r is total cost, Q is the level of output, P is a vector of the factor input prices, s is the regulatory parameter, and x is the firm's technology. This regulated cost function can be approximated by the transcendental logarithmic cost function:

$$(5.6) \quad \ln C^r = \alpha_o + \sum_i \alpha_i \ln P_i + \frac{1}{2} \sum_i \sum_j \alpha_{ij} \ln P_i \ln P_j$$

$$+ \beta_Q \ln Q + \frac{1}{2} \beta_{QQ} (\ln Q)^2 + \sum_i \beta_{Qi} \ln Q \ln P_i$$

$$+ \gamma_s \ln s + \gamma_{sQ} \ln Q \ln s + \frac{1}{2} \gamma_{ss} (\ln s)^2$$

$$+ \sum_i \gamma_{si} \ln s \ln P_i + \delta_x x + \delta_{Qx} (\ln Q) x + \delta_{sx} (\ln s) x$$

$$+ \frac{1}{2} \delta_{xx} x^2 + \sum_i \delta_{xi} (\ln P_i) x ,$$

where $i, j = k, l, f,$ and ln is the natural log operator.

The optimal cost-minimizing demand equations are obtained by differentiating (5.6) with respect to the input prices and employing Shephard's Lemma. However, unlike the unregulated case, the presence of rate of return regulation means that:

$$(5.7) \qquad \frac{\partial C^r}{\partial P_k} = \frac{\partial L^r}{\partial P_k} = q_k ,$$

$$(5.8) \qquad \frac{\partial C^r}{\partial s} = \frac{\partial L}{\partial s} = -\lambda q_k ,$$

$$(5.9) \qquad \frac{\partial C^r}{\partial P_l} = \frac{\partial L^r}{\partial P_l} = (1 - \lambda) q_l ,$$

and

$$(5.10) \qquad \frac{\partial C^r}{\partial P_f} = \frac{\partial L^r}{\partial P_f} = (1 - \lambda) q_f ,$$

where λ is the Lagrangian multiplier indicating the rate of return constraint. Of course, λ is unobservable, but following Nelson (1984), this can be overcome by rearranging the equations as follows. First, we define the derivative of equation (5.6) with respect to the log of the price of each input:

$$\left(C_k^r \right)' = \frac{\partial \ln C^r}{\partial \ln p_k}$$

$$= \alpha_k + \alpha_{kk} \ln p_k + \alpha_{kl} \ln p_l + \alpha_{kf} \ln p_f + \beta_{qk} \ln q$$

$$+ \gamma_{sk} \ln s + \delta_{xk} x ,$$

$$\left(C_s^r \right)' = \frac{\partial \ln C^r}{\partial \ln s}$$

$$= \gamma_s + \gamma_{ss} \ln s + \gamma_{sk} \ln p_k + \gamma_{sl} \ln p_l + \gamma_{sf} \ln p_f$$

$$+\gamma_{sq}\ln q+\delta_{sx}x,$$

$$\left(C_l^r\right)'=\frac{\partial\ln C^r}{\partial\ln p_l}$$

$$=\alpha_l+\alpha_{lk}\ln p_k+\alpha_{ll}\ln p_l+\alpha_{lf}\ln p_f+\beta_{ql}\ln q$$

$$+\gamma_{sl}\ln s+\delta_{xl}x,$$

and

$$\left(C_f^r\right)'=\frac{\partial\ln c}{\partial\ln p_f}$$

$$=\alpha_f+\alpha_{fk}\ln p_k+\alpha_{fl}\ln p_l+\alpha_{ff}\ln p_f+\beta_{qf}\ln q$$

$$+\gamma_{sf}\ln s+\delta_{xf}x.$$

By definition, the cost share for input i is equal to the input price times quantity, divided by total costs:

$$S_i^r=\frac{P_iq_i}{C^r}.$$

As in the unregulated case,

$$\frac{\partial\ln C^r}{\partial\ln P_i}=\frac{\partial C^r}{\partial P_i}\frac{P_i}{C^r}$$

(see equation 5.2), and by Shephard's Lemma,

$$\frac{\partial C^u}{\partial P_i}=q_i,$$

then, as a result:

$$(5.11)\quad C_k'=\frac{\partial\ln C^r}{\partial\ln P_k}=\frac{P_kq_k}{C^r}=S_k^r$$

$$=\gamma_s+\gamma_{ss}\ln s+\gamma_{sk}\ln p_k+\gamma_{sl}\ln p_l+\gamma_{sf}\ln p_f$$

$$+\gamma_{sq}\ln q+\delta_{sx}x,$$

$$(5.12)\quad C_s'=\frac{\partial\ln C^r}{\partial\ln s}=\frac{-s\lambda q_k}{C^r}=-\lambda S_k^r,$$

$$(5.13)\quad C_l'=\frac{\partial\ln C^r}{\partial\ln P_l}=\frac{P_l(1-\lambda)q_l}{C^r}=(1-\lambda)S_l^r,$$

and

$$(5.14) \quad C'_f = \frac{\partial \ln C'}{\partial \ln P_f} = \frac{P_f(1-\lambda)q_f}{C'} = (1-\lambda)S'_f.$$

Using equations (5.11) and (5.12),

$$C'_k + \frac{p_k}{s}C'_s = (1-\lambda)S'_k,$$

and dividing this into equation (5.13) reveals

$$\frac{S'_l}{S'_k} = \frac{C'_l(1-\lambda)}{\left(C'_k + \frac{p_k}{s}C'_s\right)(1-\lambda)} = \frac{C'_l}{C'_k + \frac{p_k}{s}C'_s},$$

hence,

$$(5.15) \quad S'_l = \frac{C'_l S'_k}{C'_k + \frac{p_k}{s}C'_s}$$

$$= S'_k \left(\alpha_l + \alpha_{lk} \ln p_k + \alpha_{ll} \ln p_l + \alpha_{lf} \ln p_f + \beta_{ql} \ln q \right.$$

$$+ \gamma_{sl} \ln s + \delta_{xl} x \right) \div \left\{ S'_k + \frac{P_k}{s}(\gamma_s + \gamma_{ss} \ln s + \gamma_{sk} \ln p_k \right.$$

$$+ \gamma_{sl} \ln p_l + \gamma_{sf} \ln p_f + \gamma_{sq} \ln q + \delta_{sx} x \right) \right\}.$$

Similarly,

$$\frac{S'_f}{S'_k} = \frac{C'_f(1-\lambda)}{\left(C'_k + \frac{p_k}{s}C'_s\right)(1-\lambda)} = \frac{C'_f}{C'_k + \frac{p_k}{s}C'_s},$$

hence,

$$(5.16) \quad S'_f = \frac{C'_f S'_k}{C'_k + \frac{p_k}{s}C'_s}$$

$$= S_k^r \left(\alpha_f + \alpha_{fk} \ln p_k + \alpha_{fl} \ln p_l + \alpha_{ff} \ln p_f \right.$$

$$+ \beta_{qf} \ln q + \gamma_{sf} \ln s + \delta_{xf} x \div \left\{ S_k^r + \frac{P_k}{s} (\gamma_s + \gamma_{ss} \ln s \right.$$

$$\left. + \gamma_{sk} \ln p_k + \gamma_{sl} \ln p_l + \gamma_{sf} \ln p_f + \gamma_{sq} \ln q + \delta_{sx} x \right) \}.$$

Thus, in the presence of rate of return regulation, one obtains the cost function (5.6) and the three share equations (5.11) (5.15), and (5.16), which together constitute a system of equations that can be estimated.

In addition to estimating this translog cost framework, there are also four specific hypotheses relevant to the theoretical propositions from the previous chapters that are testable: (1) was regulation effective (2) did technological change occur over the time period in question (3) did regulation impact the rate of technological change, and (4) did regulation affect the selection of factor augmenting technology? These four hypotheses can be further evaluated by dividing the sample into two parts: a pre-1975 period in which state-level rate-of-return regulation was not in effect, and a post-1975 period in which state-level rate-of-return regulation of the electric utilities in Texas was in effect.

To test the hypothesis that rate of return regulation was effective, the regulated cost function (5.6) can be compared to the unregulated cost function (5.1). This is done by performing the likelihood ratio test on the hypothesis:

$$(5.17) \quad \gamma_s = \gamma_{sq} = \gamma_{ss} = \gamma_{sk} = \gamma_{sl} = \gamma_{sf} = \delta_{sx} = 0.$$

Rejecting this hypothesis would indicate that regulation was effective.

The second hypothesis concerns the whether technological change occurred over the time period in question. This is tested by performing the likelihood ratio test on the hypothesis:

$$(5.18) \quad \delta_x = \delta_{qx} = \delta_{sx} = \delta_{xx} = \delta_{xk} = \delta_{xl} = \delta_{xf} = 0.$$

Rejecting this hypothesis would indicate that technological change did in fact occur over the time period.

The third hypothesis tests whether regulation impacted the rate of technological change. Following Granderson (1999), this can be determined by evaluating whether $\dfrac{\partial \ln C^u}{\partial x}$ differs from $\dfrac{\partial \ln C^r}{\partial x}$. From unregulated and regulated cost functions *(5.1)* and *(5.6)*, respectively, these can be defined as:

$$(5.19) \qquad \frac{\partial \ln C^u}{\partial x} = \delta_x + \delta_{qx} \ln q + \delta_{xx} x + \delta_{xk} \ln p_k + \delta_{xl} \ln p_l,$$

$$+ \delta_{xf} \ln p_f$$

and

$$(5.20) \qquad \frac{\partial \ln C^r}{\partial x} = \delta_x + \delta_{qx} \ln q + \delta_{sx} \ln s + \delta_{xx} x + \delta_{xk} \ln p_k$$

$$+ \delta_{xl} \ln p_l + \delta_{xf} \ln p_f.$$

Consequently, the impact of regulation on the rate of technical change is determined by evaluating *(5.19)* and *(5.20)* at the sample mean. If, for example, the unregulated case *(5.19)* is less than the regulated case *(5.20)*, the indication would then be that a relaxation of the regulatory constraint will lead to a decrease in costs, and thus an increase in the rate of cost-reducing technical change.

The fourth hypothesis regards the selection of factor augmenting technology. Again, following Granderson (1999), this is evaluated by comparing the change in the unregulated cost share equations with respect to technology. Hence, from equations *(5.3) (5.4)*, and *(5.5)* we evaluate:

$$(5.21) \qquad \frac{\partial S_k^u}{\partial x} = \delta_{xk},$$

$$(5.22) \qquad \frac{\partial S_l^u}{\partial x} = \delta_{xl},$$

and

$$(5.23) \qquad \frac{\partial S_f^u}{\partial x} = \delta_{xf}.$$

An indication of the firm's selection of factor augmenting technology is provided by the magnitude of each of these at the sample mean. Similarly, the regulated cost share equations can be evaluated using *(5.11) (5.15)*, and *(5.16)*:

$$(5.24) \qquad \frac{\partial S_k^r}{\partial x} = \delta_{xk},$$

$$(5.25) \quad \frac{\partial S_l^r}{\partial x} = \frac{S_k^r \delta_{xl}}{S_k^r + \frac{P_k}{s} \delta_{xl}},$$

and

$$(5.26) \quad \frac{\partial S_f^r}{\partial x} = \frac{S_k^r \delta_{xf}}{S_k^r + \frac{P_k}{s} \delta_{xf}}.$$

These can then be evaluated along with the unregulated cost share equations to get a full analysis of which factors technology augmented, as well as an indication of the impact of regulation on the selection of the factors to augment.

5.2 The Data

The sample consists of annual observations from ten privately owned electric utility companies in the state of Texas for the years 1965 to 1985. This sample comprises *all* of the private investor owned electric utilities in Texas over this time period. Omitted are the ten cooperative utilities and three municipal utilities. This data set was prepared at the Public Utility Commission of Texas (see Kwun 1987).[5] The ten electric utilities are:

- o Central Power and Light (CPL)
- o El Paso Electric (EPE)
- o Houston Lighting and Power (HLP)
- o Southwestern Electric Power (SEP)
- o Southwestern Public Service (SPS)
- o Dallas Power and Light (DPL)
- o Texas Electric Service (TES)
- o Texas Power and Light (TPL)
- o West Texas Utilities (WTU)
- o Gulf States Utilities (GSU).

Three companies, DLP, TES, and TPL, merged into one company, Texas Utilities Electric (TUE), in 1984. These observations are treated as separate companies by extrapolating from the past trend for the last two years of the data set.

To estimate the translog cost function and share equations defined in the previous section, measures are needed of output (Q), price of labor (P_l), price of fuel (P_f), price of capital (P_k), technology (x), and the allowed rate of return (s) for each of the firms.

Output (Q) for each of the ten utilities is measured as giga-watt-hours (GWH) of electricity generated by the utility plus net interchange between other utilities.

The price of labor (P_l) is defined as the average annual wage, determined by dividing labor expenditures by the quantity of labor (q_l).[6] Labor expenditures are calculated as the sum of wages, salaries, and employee pensions and benefits. Quantity of labor is calculated as the number of full-time employees plus one-half times the number of part time employees.

The price of fuels (P_f) is calculated as the weighted average of the price of all types of fuel used by each utility. Fuel quantity (q_f) is obtained by dividing fuel expenditures by the average price of fuels (P_f). Fuel expenditures are calculated as expenditures from all types of fuel used for generation.

The price of capital (P_k) is measured as capital expenditures divided by the quantity of capital (q_k). Capital expenditures are calculated as the long-term interest payments multiplied by the ratio of total capital to long term debts, plus depreciation expenses. Quantity of capital is measured as the deflated stock of capital (see Kwun 1987 for further details).

Following Nelson (1984) and Granderson (1999), technology (x) is measured as the time trend. Interpretations of this measure require caution, but using the time-trend does create a very general indicator of not only factor improvements, but also of other sources of technical change such as organizational changes and learning by doing.

The allowed rate of return (s) is calculated as the utilities' operating income divided by the net electric utility plant deflated by the Handy-Whitman index (see Kwun 1987 for further details). Finally, total cost

Table 5.1 Means and Standard Deviations

Symbol	Variable	Mean	Standard Deviation
Q	Output (GWH)	15147.91	12583.07
q_l	Quantity of Labor (FT + 0.5*PT workers)	2752.41	1930.19
q_f	Quantity of Fuel (1000 MMBTU)	166725.55	138107.59
q_k	Quantity of Capital (10000$)	171105.25	114149.89
P_l	Price of Labor (1000$ / worker)	12.956	8.035
P_f	Price of Fuel ($ / MMBTU)	1.079	1.007
P_k	Price of Capital (0.1 percent)	0.775	0.248
s	Allowed Rate of Return (percent)	7.376	1.108
C	Total Cost	402376.85	518567.30

(C) is the sum of capital, labor, and fuel costs. Table 5.1 provides a listing of the mean and standard deviation of all the variables used in the model.

5.3 Estimation and Results

Since statewide use of rate-of-return regulation did not begin until 1975, the data must be estimated with two separate systems of equations. For the pre-1975 period, the system includes the cost function (5.1) and the share equations (5.3) (5.4), and (5.5). However, since this function is homogenous of degree one in prices, the following restrictions must be imposed:

$$(5.27) \quad \sum_i \alpha_i = 1 \quad \text{and} \quad \sum_i \alpha_{ij} = \sum_j \alpha_{ij} = \sum_i \beta_{Qi} = 0.$$

With these restrictions, one of the three share equations must be dropped to avoid singularity of the covariance matrix. Fortunately, Zellner's nonlinear method of iterative seeming unrelated regression (NLITSUR) provides coefficient estimates which are invariant with respect to the omitted share equation (see Zellner 1962, 1963). We deleted the labor share equation (5.4).

For the post-1975 period, the system includes the rate-of-return adjusted cost function (5.6) and share equations (5.11) (5.15), and (5.16). Unlike the previous cost function and share equations, this function is *not* homogenous of degree one in prices (see Cowing 1982), thus we must estimate all three share equations without the homogeneity restrictions imposed *a priori*.

In both systems of equations autocorrelation is present. To correct for this, the method advocated by Berndt and Savin (1975) for autocorrelated translog cost functions is used (for a similar use see Jewell, O'Brien, and Seldon, forthcoming).[7] For the unregulated model, the error terms are:

$$(5.28) \quad u_{c,t} = \rho_{c,c} u_{c,t-1} + \rho_{sk,c} u_{sk,t-1} + \rho_{sf,c} u_{sf,t-1} + v_{c,t},$$

$$(5.29) \quad u_{sk,t} = \rho_{c,sk} u_{c,t-1} + \rho_{sk,sk} u_{sk,t-1} + \rho_{sf,sk} u_{sf,t-1} + v_{sk,t},$$

and

$$(5.30) \quad u_{sf,t} = \rho_{c,sf} u_{c,t-1} + \rho_{sk,sf} u_{sk,t-1} + \rho_{sf,sf} u_{sf,t-1} + v_{sf,t},$$

where $v_{i,t}$ are well-behaved error terms.

The lagged error term in each of these is defined as

$$u_{i,t-1} = \alpha_i + \alpha_{ij} \ln P_{j,t-1} + \beta_{Qi,t-1} \ln Q_{t-1} + \delta_{xi} x_{t-1}.$$

The lagged error for the cost function ($u_{c,t}$) is similarly estimated.

In the regulated model, all share equations are estimated and the homogeneity restrictions are not imposed. Thus, for the regulated model, the error terms are:

$$(5.31) \quad u_{c,t} = \rho_{c,c} u_{c,t-1} + \rho_{sk,c} u_{sk,t-1} + \rho_{sf,c} u_{sf,t-1} + \rho_{sl,c} u_{sl,t-1} + v_{c,t},$$

$$(5.32) \quad u_{sk,t} = \rho_{c,sk} u_{c,t-1} + \rho_{sk,sk} u_{sk,t-1} + \rho_{sf,sk} u_{sf,t-1} + \rho_{sl,sk} u_{sl,t-1}$$
$$+ v_{sk,t},$$

$$(5.33) \quad u_{sf,t} = \rho_{c,sf} u_{c,t-1} + \rho_{sk,sf} u_{sk,t-1} + \rho_{sf,sf} u_{sf,t-1} + \rho_{sl,sf} u_{sl,t-1}$$
$$+ v_{sf,t},$$

and

$$(5.34) \quad u_{sl,t} = \rho_{c,sl} u_{c,t-1} + \rho_{sk,sl} u_{sk,t-1} + \rho_{sf,sl} u_{sf,t-1} + \rho_{sl,sl} u_{sl,t-1}$$
$$+ v_{sl,t}.$$

Much like the unregulated case before, the lagged error term in each of these is defined as

$$u_{i,t-1} = S^r_{i,t-1},$$

where $S^r_{i,t-1}$ follows from equations (5.11) (5.15), and (5.16). The lagged error for the cost function ($u_{c,t}$) is similarly estimated.

Table 5.2 provides the autocorrelation-corrected NLITSUR parameter estimates for the pre-1975 and post-1975 cost functions and share equations. Given the lagged structure of the error term, only observations in which lagged data was available were used. In addition, because of convergence problems in the post-1975 model, the criteria for judging singularity was changed in the post model from 1E-9 to 1E-7. As should be expected, each of the share equations from both the unregulated and regulated models is nonnegative when calculated at the variable means.[8] Parameter estimates from a pooled model which includes the full time period (1965-1985) are provided in the appendix to this chapter.

There are four specific hypotheses that are testable from the translog cost framework: (1) was regulation effective (2) did technological change occur over the time period in question (3) did regulation impact the rate of technological change, and (4) does regulation affect the selection of factor augmenting technology?

Table 5.2 NLITSUR Parameter Estimates of the Translog Cost Function and Share Equations

Coefficient	Pre Model (1965-1974)	Post Model (1975-1985)
α_0	11.8439 (2.010)*	0.6804 (0.893)
α_l	-4.0460 (1.104)*	0.1321 (0.013)*
α_f	0.2919 (0.038)*	0.0105 (0.013)
α_k	0.9545 (0.054)*	0.8567 (0.000)*
α_{ll}	-0.1107 (0.111)	0.0100 (0.002)*
α_{ff}	0.2023 (0.010)*	0.2198 (0.008)*
α_{kk}	0.2714 (0.023)*	0.1832 (0.008)*
α_{lf}	-0.0564 (0.008)*	-0.0232 (0.004)*
α_{fk}	-0.1562 (0.011)*	-0.1961 (0.007)*
α_{kl}	-0.0906 (0.010)*	0.0132 (0.003)*
β_Q	0.0529 (0.233)	1.6227 (0.181)*
β_{QQ}	0.0023 (0.003)	-0.0835 (0.020)*
β_{Ql}	0.4575 (0.136)*	-0.0129 (0.000)*
β_{Qf}	0.0257 (0.004)*	0.0545 (0.000)*
β_{Qk}	-0.0246 (0.005)*	-0.0414 (0.000)*
γ_s	–	0.0011 (0.001)
γ_{Sq}	–	0.0001 (0.000)*
γ_{ss}	–	0.0007 (0.000)

Continued on next page.

Table 5.2 (Continued)

Coefficient	Pre Model (1965-1974)	Post Model (1975-1985)
γ_{sl}	–	0.0006 (0.000)
γ_{sf}	–	-0.0012 (0.000)*
γ_{sk}	–	0.0002 (0.000)
δ_x	0.0580 (0.077)	-0.0568 (0.035)
δ_{Qx}	-0.0100 (0.007)	0.0087 (0.003)*
δ_{sx}	–	-0.0001 (0.000)*
δ_{xx}	0.0027 (0.004)	-0.0019 (0.002)
δ_{xl}	0.0056 (0.027)	0.0042 (0.001)*
δ_{xf}	0.0047 (0.001)*	-0.0027 (0.001)*
δ_{xk}	0.0002 (0.002)	-0.0015 (0.001)*
$\rho_{c,c}$	0.0000 (0.000)	0.0000 (0.000)*
$\rho_{sk,c}$	0.0056 (0.017)	-0.0006 (0.000)*
$\rho_{sf,c}$	0.0049 (0.016)	0.2625 (0.000)*
$\rho_{sl,c}$	–	1.1143 (0.000)*
$\rho_{c,sk}$	0.0000 (0.000)	-0.0000 (0.000)*
$\rho_{sk,sk}$	0.0048 (0.004)	0.0001 (0.000)*
$\rho_{sf,sk}$	0.0046 (0.004)	-0.1080 (0.000)*
$\rho_{sl,sk}$	–	-0.4580 (0.000)*
$\rho_{c,sf}$	-0.0000 (0.000)	0.0000 (0.000)*
$\rho_{sk,sf}$	-0.0082 (0.006)	-0.0001 (0.000)*

Continued on next page.

Table 5.2 (Continued)

Coefficient	Pre Model (1965-1974)	Post Model (1975-1985)
$\rho_{sf,sf}$	-0.0078 (0.005)	0.0257 (0.000)*
$\rho_{sl,sf}$	–	0.1091 (0.000)*
$\rho_{c,sl}$	–	0.0000 (0.000)*
$\rho_{sk,sl}$	–	-0.0001 (0.000)*
$\rho_{sf,sl}$	–	0.0840 (0.000)*
$\rho_{sl,sl}$	–	0.3562 (0.000)*
R^2 (lnC)	0.9887	0.9950
R^2 (S_k)	0.8033	0.8674
R^2 (S_f)	0.8946	0.8765
R^2 (S_l)	–	0.4689

Note: Asymptotic *t*-statistic in parentheses. * indicates significance at the .05 level.

The first hypothesis is tested by using the likelihood ratio test on restriction (5.17). In the post-1975 model, the χ^2 test statistic is 97.266, and thus the hypothesis that regulation was ineffective is rejected at the 1% level. This hypothesis is not applicable to the pre-1975 period because rate-of-return regulation was not imposed during this period.

The second hypothesis is tested by using the likelihood ratio test on restriction (5.18). In the pre-1975 model, the resulting χ^2 test statistic is 40.442, and thus the hypothesis of no technical change is easily rejected at the 1% level. In the post-1975 model, the χ^2 test statistic is 49.421, and thus the hypothesis of no technical change is again rejected at the 1% level.

The third hypothesis considers whether the presence of rate-of-return regulation affected the rate of technological change. Accordingly, this is judged by evaluating at the sample mean from (5.19) for the pre-model, and (5.20) for the post-model. For the pre-model we find that

$\partial \ln C^u / \partial x = -0.0124$, which is statistically significant at the .01 level (t-stat $= -18.043$). This indicates that due to technological progress, costs on average decreased by 1.24% per year before the statewide implementation of rate of return regulation. For the post-model, we find that $\partial \ln C^r / \partial x = 0.0072$, which again is statistically significant at the .01 level (t-stat $= 4.632$). This indicates that due to technological regress, costs increased by an average of 0.72% per year after the statewide implementation of rate of return regulation.

The fourth hypothesis concerns whether regulation affects the choice of factor augmenting technology. This is evaluated by estimating the change in the share equations with respect to technology. For the pre-model, this done by calculating equations (*5.21*) (*5.22*) and (*5.23*) at the sample mean. For the post-model, this is done by calculating equations (*5.24*) (*5.25*) and (*5.26*) at the sample mean.

In the period before the implementation of statewide rate of return regulation, we find that:

$$\frac{\partial S_k^u}{\partial x} = 0.0002,$$

$$\frac{\partial S_l^u}{\partial x} = 0.0056,$$

and

$$\frac{\partial S_f^u}{\partial x} = 0.0047.$$

In the pre-model, the input share of capital increased by 0.02% per year with respect to technology, although this is not statistically significant at the .05 level (t-stat $= 0.878$). At the same time, the input share of labor increased by 0.56% per year with respect to technology, while the input share of fuel decreased by 0.47% per year. Both of these derivatives are statistically significant at the .05 level (t-stat $= 1.996$ and 31.167 respectively). This implies that during this period before the implementation of rate of return regulation, the utility firms adopted technology that augmented noncapital (labor and fuel) more than capital, and technology that augmented labor more than fuel:

$$\frac{\partial S_l^u}{\partial x} > \frac{\partial S_f^u}{\partial x} > \frac{\partial S_k^u}{\partial x}.$$

It should be noted, however, that the magnitudes of these derivatives are quite small.

In the period after the implementation of statewide rate of return regulation we find that:

$$\frac{\partial S_k^r}{\partial x} = -0.0015,$$

$$\frac{\partial S_l^r}{\partial x} = 0.0035,$$

and

$$\frac{\partial S_f^r}{\partial x} = -0.0031.$$

Thus, the input share of labor increased by 0.35% per year with respect to technology, while the input shares of capital and fuel decreased by 0.25% and 0.31% per year respectively. All of these are statistically significant at the .01 level (t-stat = -21.575, 33.828, and -27.890 respectively). This implies that after the implementation of rate of return regulation, the utility firms adopted technology that augmented labor more than capital, and capital more than fuel:

$$\frac{\partial S_l^r}{\partial x} > \frac{\partial S_k^r}{\partial x} > \frac{\partial S_f^r}{\partial x}.$$

This is supportive of the theoretical propositions of Smith (1974), and proposition 3.8 from Chapter 3.

These results also indicate that firms adopted more capital augmenting technology before the statewide implementation of rate of return regulation than after:

$$\frac{\partial S_k^u}{\partial x} > \frac{\partial S_k^r}{\partial x}.$$

This result supports proposition 3.8 from chapter 3, and replicates the findings of Granderson (1999). The same is true for labor and fuel: after the implementation of regulation, firms adopted less labor and fuel augmenting technology than before:

Table 5.3 Summary of Empirical Results

Hypothesis	Pre-1975 Model (1965-1974)	Post-1975 Model (1975-1985)
H1: Regulation Ineffective	n/a	reject
H2: No Technological Change	reject	reject
H3: Regulation and the Rate of Technical Change	$\dfrac{\partial \ln C^u}{\partial x} < 0$	$\dfrac{\partial \ln C^r}{\partial x} > 0$
H4: Selection of Factor Augmenting Technology	$\dfrac{\partial S_l^u}{\partial x} > \dfrac{\partial S_f^u}{\partial x} > \dfrac{\partial S_k^u}{\partial x}$	$\dfrac{\partial S_l^r}{\partial x} > \dfrac{\partial S_k^r}{\partial x} > \dfrac{\partial S_f^r}{\partial x}$
		$\dfrac{\partial S_k^u}{\partial x} > \dfrac{\partial S_k^r}{\partial x}$

$$\frac{\partial S_l^u}{\partial x} > \frac{\partial S_l^r}{\partial x} \quad \text{and} \quad \frac{\partial S_l^u}{\partial x} > \frac{\partial S_l^r}{\partial x}.$$

Again, the magnitudes of these derivatives are quite small, indicating that the overall effect of regulation was quite small.

5.4 Summary

The principal results from this chapter are summarized in Table 5.3. The first hypothesis (H1) concerns the general effectiveness of rate of return regulation on the firm. The results indicate that the presence of rate-of-return regulation did matter.

The second hypothesis (H2) concerns the presence of technological change. The results indicate that a significant degree of technological change occurred, but tell us nothing about the direction or magnitude of the technological change.

The third hypothesis (H3) tests the directional effect of regulation on the rate of technological change. The results indicate that before the statewide implementation of rate return regulation, costs were significantly lowered by technological progress. After the statewide implementation, however, costs increased due to technological *regress*. This conclusion concurs with the principal theoretical findings from Chapter 3.

Finally, the fourth hypothesis (H4) concerns whether regulation affected the selection of factor augmenting technology. Although the results indicate that the substantive impact was quite small, they do generally support the conclusion by Smith (1975) that firms will adopt less capital augmenting technology once regulation is imposed. The results also indicate that firms used less labor and fuel augmenting technology once regulation was imposed.

Notes

[1] The numerous and high profile legal problems of Southwestern Bell Telephone Co. also appear to have been a key contributor of the demand for utility regulation (see Adams 1976).

[2] Smith's model is presented in section 2.3.4 of Chapter 2.

[3] The cost function is estimated instead of the production function because the production function is a minimum of two functions and thus not differentiable (see Nelson 1984). This does not, however, present a problem because of the well known dual relationship between costs and production.

[4] Of course, the cost share for input i is by definition equal to the input price times quantity, divided by total costs:

$$S_i^u = \frac{P_i q_i}{C^u}.$$

[5] Kumbhakar (1996) also uses this data set for a parametric measurement of efficiency using a generalized profit function.

[6] Since expenditures for an input are equal to the price of the input times its quantity (E = P·Q), then it must also be true that P = E÷Q.

[7] By comparison, the Brendt-Savin method is superior to other methods such as ITSUR combined with Cochrane-Orcutt, because the model coefficients are estimated simultaneously with the autocorrelation coefficients.

[8] Specifically, using the variables at their mean, the unregulated capital and fuel share equations are equal to 0.4068 and 0.5017, respectively (t-stat = 29.9173 and 37.7863). Likewise, the regulated capital, labor, and fuel share equations are equal to 0.0038, 0.0686, and 0.4617, respectively (t-stat = 1.5975, 7.9118, and 56.8634).

Appendix: Pooled Model

This appendix provides parameter estimates from a pooled model which encompasses the time period 1965-1985. Although this pooled model has the advantage of greater efficiency in the parameter estimates, it suffers because the homogeneity restrictions (*5.27*) cannot be imposed for the 1965-1974 period. As a result, the pooled model is estimated using the translog cost function and all three share equations as if the function was not homogenous in input prices for the entire period.

Table A5.1 provides the autocorrelation-corrected NLITSUR parameter estimates for the cost function and share equations from the pooled model. From these estimates, four specific hypotheses are tested: (1) was regulation effective, (2) did technological change occur over the time period in question, (3) did regulation impact the rate of technological change, and (4) did regulation effect the selection of factor augmenting technology?

The first hypothesis is tested by using the likelihood ratio test on restriction (*5.17*). The pooled model χ^2 test statistic is 8.714, which is not statistically significant. Therefore, we fail to reject the hypothesis that regulation was ineffective in the pooled model.[1]

The second hypothesis is tested by using the likelihood ratio test on restriction (*5.18*). The resulting χ^2 test statistic is 107.271, and is significant at the 1% level. Consequently, the hypothesis of no technical change is rejected.

The third hypothesis considers whether the presence of rate-of-return regulation affected the rate of technological change. This is judged by evaluating at the sample mean from equation (*5.20*). We find that $\partial \ln C^r / \partial x = -0.0036$, which is statistically significant at the .01 level (*t*-stat = -35.033). This indicates that over the entire sample period (1965 to 1985), costs decreased by an average of 0.36% per year because of technological progress.

The fourth hypothesis concerns whether regulation affects the choice of factor augmenting technology. This is evaluated by estimating the change in the share equations with respect to technology. Here we find that:

Table A5.1 NLITSUR Parameter Estimates from the Pooled Model

Coefficient	Pooled Model (1965-1985)	Coefficient	Pooled Model Continued
α_0	-1.0594 (0.527)*	δ_{xx}	-0.0023 (0.000)*
α_l	-0.1087 (0.023)	δ_{xl}	-0.0040 (0.001)*
α_f	0.0700 (0.027)*	δ_{xf}	-0.0008 (0.001)
α_k	1.0397 (0.032)*	δ_{xk}	0.0048 (0.001)*
α_{ll}	0.1225 (0.005)*	$\rho_{c,c}$	-0.0000 (0.000)
α_{ff}	0.2140 (0.006)*	$\rho_{sk,c}$	-0.0000 (0.000)
α_{kk}	0.2672 (0.009)*	$\rho_{sf,c}$	-0.0003 (0.000)
α_{lf}	-0.0348 (0.003)*	$\rho_{sl,c}$	-0.0017 (0.003)
α_{fk}	-0.1791 (0.006)*	$\rho_{c,sk}$	0.0000 (0.000)
α_{kl}	-0.0877 (0.006)*	$\rho_{sk,sk}$	-0.0001 (0.000)*
β_Q	2.2430 (0.122)*	$\rho_{sf,sk}$	-0.0007 (0.000)*
β_{QQ}	-0.1669 (0.014)*	$\rho_{sl,sk}$	-0.0049 (0.001)*
β_{Ql}	-0.0070 (0.002)*	$\rho_{c,sf}$	-0.0000 (0.000)
β_{Qf}	0.0466 (0.003)*	$\rho_{sk,sf}$	0.0000 (0.000)*
β_{Qk}	-0.0397 (0.003)*	$\rho_{sf,sf}$	0.0007 (0.000)*
γ_s	0.0076 (0.001)*	$\rho_{sl,sf}$	0.0050 (0.001)*
γ_{sQ}	-0.0008 (0.000)*	$\rho_{c,sl}$	-0.0000 (0.000)
γ_{ss}	0.0023 (0.000)*	$\rho_{sk,sl}$	0.0000 (0.000)
γ_{sl}	-0.0008 (0.000)*	$\rho_{sf,sl}$	-0.0000 (0.000)

Continued on next page.

Table A5.1 (Continued)

γ_{sf}	-0.0015 (0.000)*	$\rho_{sl,sl}$	-0.0002 (0.001)
γ_{sk}	0.0024 (0.000)*	R^2 $(\ln C)$	0.9968
δ_x	-0.1453 (0.011)*	R^2 (S_k)	0.9637
δ_{Qx}	0.0192 (0.001)*	R^2 (S_f)	0.9767
δ_{sx}	0.0000 (0.000)*	R^2 (S_l)	0.8628

Note: Asymptotic t-statistic in parentheses. * indicates significance at the .05 level.

$$\frac{\partial S_k}{\partial x} = 0.0048 , \quad \frac{\partial S_l}{\partial x} = -0.0043 ,$$

and

$$\frac{\partial S_f}{\partial x} = -0.0008 .$$

The input share of capital increased by 0.48% per year with respect to technology, while the input share of labor and fuel decreased by 0.43% and 0.08% per year respectively. All of these are statistically significant at the .01 level (t-stat = 97.072, −58.544, and −10.315 respectively). Over the entire period of the sample, utility firms tended to adopt technology which augmented capital more than fuel, and fuel more than labor:

$$\frac{\partial S_k}{\partial x} > \frac{\partial S_f}{\partial x} > \frac{\partial S_l}{\partial x} .$$

Note

[1] As previously noted, the pooled model may be biased because the homogeneity restrictions cannot be imposed for the years 1965-1975.

6 Implications and Conclusions

The guiding research question of this study concerns the impact of government regulation on the innovativeness of firms with market power. The investigation began in with an historical overview of the development of regulation in general and rate-of-return regulation in particular, and with a broad overview of the process and players from a typical rate hearing. This was followed by a discussion of the theoretical model of rate-of-return regulation put forth by Averch and Johnson (1962), as well as many of the key extensions and criticisms of the model. A variation of the Averch-Johnson model which includes technology as a third factor of production then was outlined. With this model, rate-of-return regulation was shown to decrease the amount of technology relative to capital and to increase the use of noncapital augmenting technology relative to capital augmenting technology. Hence, the Averch-Johnson effect not only causes an overuse of capital relative to labor, but also an overuse of capital relative to technology. This logic could in fact be extended to any number of production inputs.

Following this, a two-stage Nash equilibrium duopoly model was considered in which each of the firms is subject to rate-of-return regulation. From this two-stage game, it was discovered that although regulation reduces R&D, the reduction can be partially mitigated by participation in joint research ventures. This outcome is in part a result of the Averch-Johnson effect and in part a result of appropriability of R&D benefits to the individual firm. With technology as an additional factor of production we see that not only does rate-of-return regulation cause an overuse of capital relative to labor, but also an overuse of capital relative to technology. Hence, as regulation increases, expenditures on R&D decrease. On the other hand, when research spillovers are sufficiently high, competing firms have little incentive to enact costly R&D investments since the benefits from doing so will not be exclusive to themselves. Once engaged in an R&D joint venture, however, this incentive changes and R&D expenditures increase.

Consequently, although rate-of-return regulation reduces R&D, this reduction can be partially mitigated by participation in a joint research venture.

Several of these key theoretical propositions were tested empirically using a sample of ten electric utility firms from the state of Texas covering the years 1965 to 1985. The results indicate that the use of rate-of-return regulation decreased the rate of technological innovation and caused firms to adopt less capital augmenting technology than before. This conclusion supports the key theoretically propositions from the augmented Averch-Johnson model.

6.1 Welfare Implications

An additional question relates to the welfare implications of these results. Whereas rate-of-return regulation has the unintended consequence of decreasing the firm's expenditure on research and development, this result alone indicates little from a social welfare perspective. While increasing R&D may increase profit for the firm, it is unclear whether it benefits consumers in the same way. A social welfare analysis is useful because it provides an important efficiency standard by which to gauge the overall effect of rate-of-return regulation on innovation.

Social welfare (SW) is defined as total industry consumer surplus (CS) plus total industry profit ($TR - TC$). In Chapter 3, a profit function for the unregulated i^{th} firm which includes technology as a third factor of production was defined as:

(3.3) $\quad \Pi_i = P(Q(q_i(x_i,k_i,l_i)))q_i(x_i,k_i,l_i) - rk_i - wl_i - \gamma x_i .$

Consequently, in symmetric equilibrium ($k_i = k_j$, $l_i = l_j$, $x_i = x_j$), social welfare is

(6.1) $\quad SW = \int_0^Q P(Q(x_i,k_i,l_i))dQ - rk_i - wl_i - \gamma x_i .$ [1]

As before, the imposition of rate-of-return regulation means that for the ith firm

(6.2) $\quad P(Q(x_i,k_i,l_i))Q(x_i,k_i,l_i) - sk_i - wl_i - \gamma x_i = 0 ,$

where s is the allowed cost of capital. Therefore, the regulated social welfare function is:

$$(6.3) \quad SW^r = \int_0^Q P(Q(x_i, k_i, l_i)) dQ - rk_i - wl_i - \gamma x_i$$

$$- \lambda [P(Q(x_i, k_i, l_i)) Q(x_i, k_i, l_i) - sk_i - wl_i - \gamma x_i],$$

where $0 \leq \lambda \leq 1$, and $s > r$. Recall from Chapter 3, that λ reflects the degree and effectiveness of regulation. With perfect rate-of-return regulation, λ will equal one, but with perfectly ineffective regulation, λ will equal zero. Moreover, as λ approaches one, regulation becomes more effective, but as λ falls towards zero, regulation becomes less effective.

As a result, the first order condition with respect to R&D is:

$$\frac{\partial SW^r}{\partial x_i} = -\gamma(1-\lambda) + P(1-\lambda)\frac{\partial Q}{\partial x_i} - \lambda Q \frac{dP}{dQ}\frac{\partial Q}{\partial x_i} = 0.$$

This can then be solved for the marginal product of R&D:

$$(6.4) \quad MP_x = \frac{\partial Q}{\partial x_i} = \frac{-\gamma(1-\lambda)}{-P(1-\lambda) + \lambda Q \dfrac{dP}{dQ}}$$

$$= \frac{\gamma}{P - \left(\dfrac{\lambda}{1-\lambda}\right) Q \dfrac{dP}{dQ}},$$

where $dP/dQ < 0$. The implication from this is that the social welfare maximizing marginal product of R&D (MP_x) is inversely related to regulation (λ). This occurs because as λ approaches 1 (perfect regulation), then the marginal product of R&D falls, and in the limit is equal zero:

$$\underset{\lambda \to 1}{Lim} \left[\frac{\gamma}{P - \dfrac{\lambda}{1-\lambda} Q \dfrac{dP}{dQ}} \right] = 0,$$

precisely because

$$\underset{\lambda \to 1}{Lim}\left[-\frac{\lambda}{1-\lambda}Q\frac{dP}{dQ} \right] = +\infty .$$

On the other hand, as λ falls to zero (no regulation), the marginal product of R&D increases, and in the limit is equal to the ratio of γ to P:

$$\underset{\lambda \to 0^+}{Lim}\left[\frac{\gamma}{P-\dfrac{\lambda}{1-\lambda}Q\dfrac{dP}{dQ}} \right] = \frac{\gamma}{P}$$

where γ, $P > 0$.   Hence, as rate-of-return regulation increases (decreases), *ceteris paribus*, the marginal product of R&D decreases (increases).

This conclusion follows from the most general case in which technology is simply a production input of some unknown form. The results are therefore fully general, but also unspecific. For illustrative purposes, the next subsection explores an example where production is given the more specific Cobb-Douglas form. This is useful because it gives clear and tangible results regarding the implications of rate-of-return regulation for social welfare.

Cobb-Douglas Example

Suppose production follows the classic Cobb-Douglas form:

$$Q = x_i k_i^\alpha l_i^{1-\alpha} ,$$

where $0 < \alpha < 1$. Social welfare (SW) is defined as total industry consumer surplus (CS) plus total industry profit $(TR - TC)$. Thus, in symmetric equilibrium $(k_i = k_j,\ l_i = l_j,\ x_i = x_j)$ social welfare for the unregulated industry is:

$$(6.5)\qquad SW = \int_0^Q P(Q(x_i,k_i,l_i))dQ - rk_i - wl_i - \gamma x_i .$$

The imposition of rate-of-return regulation means that

$$(6.6)\ P(Q(x_i,k_i,l_i))x_i k_i^\alpha l_i^{1-\alpha} - sk_i - wl_i - \gamma x_i = 0 ,$$

where s is the allowed cost of capital. Therefore, the regulated social welfare function is equal to:

(6.7) $\quad SW^r = \int_0^Q P(Q(x_i,k_i,l_i))dQ - rk_i - wl_i - \gamma x_i$

$$- \lambda \left[P(Q(x_i,k_i,l_i))x_i k_i^\alpha l_i^{1-\alpha} - sk_i - wl_i - \gamma x_i \right]$$

where $0 \le \lambda \le 1$, and $s > r$. The resulting first order condition with respect to R&D is:

$$\frac{\partial SW^r}{\partial x_i} = -\gamma(1-\lambda) - k_i^\alpha l_i^{1-\alpha} P\lambda - k_i^{2\alpha} l_i^{2-2\alpha} x_i \frac{dP}{dQ}\lambda = 0.$$

This can then be solved for R&D:

(6.8) $\quad x_i = -\dfrac{\gamma(1-\lambda) + k_i^\alpha l_i^{1-\alpha} P\lambda}{k_i^{2\alpha} l_i^{2-2\alpha} \dfrac{dP}{dQ}\lambda}$

$$= \left(\frac{1-\lambda}{\lambda}\right)\left[\frac{\gamma}{-\dfrac{dP}{dQ}\left(k_i^\alpha l_i^{1-\alpha}\right)^2}\right] + \left(\frac{P}{-\dfrac{dP}{dQ}k_i^\alpha l_i^{1-\alpha}}\right) \ge 0,$$

because $dP / dQ < 0$. The implication is that the social welfare maximizing level of R&D (x_i) is inversely related to regulation (λ). This follows because as λ approaches one (perfect regulation), R&D falls and in the limit:

$$\underset{\lambda \to 1}{Lim}\left\{\left(\frac{1-\lambda}{\lambda}\right)\left[\frac{\gamma}{-\dfrac{dP}{dQ}\left(k_i^\alpha l_i^{1-\alpha}\right)^2}\right] + \left(\frac{P}{-\dfrac{dP}{dQ}k_i^\alpha l_i^{1-\alpha}}\right)\right\}$$

$$= \frac{P}{-\dfrac{dP}{dQ}k_i^\alpha l_i^{1-\alpha}} > 0,$$

precisely because

$$\underset{\lambda \to 1}{Lim}\left(\frac{1-\lambda}{\lambda}\right) = 0.$$

On the other hand, as λ approaches zero (no regulation), R&D increases, and in the limit:

$$\underset{\lambda \to 0^+}{Lim}\left\{\left(\frac{1-\lambda}{\lambda}\right)\left[\frac{\gamma}{-\frac{dP}{dQ}\left(k_i^\alpha l_i^{1-\alpha}\right)^2}\right]+\left(\frac{P}{-\frac{dP}{dQ}k_i^\alpha l_i^{1-\alpha}}\right)\right\} = +\infty.$$

Consequently, the social welfare maximizing level of R&D is inversely related to the degree of rate-of-return regulation: as regulation increases (decreases), R&D decreases (increases). Indeed, R&D is greatest, from a social welfare perspective, when there is no regulation. The intuition behind this result is that the reduction in the quantity of technological input resulting from the imposition of rate-of-return regulation, *decreases* social welfare more than the reduction of profits resulting from rate-of-return regulation *increases* social welfare. Hence, the imposition of rate-of-return regulation has the net effect of decreasing social welfare precisely because the damage from less R&D is greater than the benefit of lower profits.

6.2 Next Step

We find that using rate-of-return regulation does have the unintended consequence of decreasing the firm's expenditures on R&D. Rate-of-return regulation is shown to cause the firm to overuse capital relative to both labor and R&D, and to overuse labor augmenting R&D relative to capital augmenting R&D. This decrease can be partially mitigated by participation in research joint ventures, however. Rate-of-return regulation also has the effect of moving R&D below the social welfare maximizing level: from a Cobb-Douglas example, R&D is shown to be greatest from a social welfare perspective when regulation is nonexistent.

Fortunately, over the last decade, the incidence of rate-of-return regulation as a policy instrument has been declining. On the federal level, during the 1980s the FERC moved away from rate-of-return regulation in the gas industry, to various forms of incentive regulation. Likewise, the FCC began using price-caps instead of rate-of-return regulation to regulate AT&T in 1989. In the state of Texas, the Public Utility Commission of Texas has begun a process of gradual decontrol of the telephone industry, and in the summer of 1999 passed legislation which will deregulate the electric industry. It is likely that this trend away from rate-of-return regulation to price caps and other forms of incentive regulation, will continue in the coming years.

These recent events and trends do not diminish the importance of the research conducted here, however. In part, this is because even purely historical questions are relevant to current policy problems and issues. Learning from the failures of previous policies is crucial precisely because if the failures of these past methods are not properly identified, their replacements are just as likely to repeat their failures. Only after the failures of these previous methods have been properly identified can the new methods be tested for improvement. Comparatively speaking, if the key faults of rate-of-return regulation are never properly identified and recognized, then any assessment of price caps or some other form of incentive regulation will be inherently flawed.

The implication here is that the key weakness of rate-of-return regulation is its negative impact on technological innovation. What remains unanswered is what effect these new forms of regulation will have on technological innovation. Will they, for example, encourage firms to engage in more innovation than occurred under rate-of-return regulation? Will they cause regulated monopolies to engage in more innovation then unregulated monopolies? Will social welfare be greater with more regulation? And how will these outcomes compare to those seen under rate-of-return regulation?

These important questions constitute the next critical steps in furthering the research paradigm which underlies this study. This paradigm places importance not solely on price, quantity, and market structure, but also on the long run affects resulting from technology and innovation. In the words of Joseph Schumpeter:

> However it is still competition within a rigid pattern of invariant conditions, methods of production and forms of industrial organization in particular, that practically monopolizes attention. But in capitalist reality as distinguished from its textbook picture,

it is not that kind of competition which counts but competition from the new commodity, the new technology, the new source of supply, the new type of organization (the largest scale unit of control for example) – competition which commands a decisive cost or quality advantage and which strikes not at the margins of the profits and the outputs of the existing firms but at their foundations and their very lives. This kind of competition is as much more effective than the other as a bombardment is in comparison with forcing a door, and so much more important that it becomes a matter of comparative indifference whether competition in the ordinary sense functions more or less promptly; the powerful lever that in the long run expands output and brings down prices is in any case made of other stuff (Schumpeter, 1950, 84-85).

Note

[1] This follows because the area under demand and up to output Q is also defined as consumer surplus plus total revenue: $\int_0^Q P(Q(x_i, k_i, l_i))dQ = CS + TR$.

Bibliography

Adams, Don. 1976. Utility Regulation: A Public Demand. *Baylor Law Review* 28 (4, Fall):773-76.

Alger, Dan, and Michael Toman. 1990. Market-Based Regulation of Natural Gas Pipelines. *Journal of Regulatory Economics* 2:263-80.

Arrow, Kenneth J. 1962. The Economics of Learning by Doing. *Review of Economic Studies* 29 (June):155-73.

Averch, Harvey, and Leland L. Johnson. 1962. Behavior of the Firm Under Regulatory Constraint. *American Economic Review* 52 (5, December):1052-69.

Bailey, Elizabeth E., and Roger D. Coleman. 1971. The Effect of Lagged Regulation in an Averch-Johnson Model. *Bell Journal of Economics and Management Science* 2 (1):278-92.

Bailey, Elizabeth E., and John C. Malone. 1970. Resource Allocation and the Regulated Firm. *Bell Journal of Economics and Management Science* 1 (1, Spring):129-42.

Barro, Robert J., and Xavier Sala-i-Martin. 1995. *Economic Growth*. New York: McGraw-Hill.

Baumol, William J., and Alvin K. Klevorick. 1970. Input Choices and Rate-of-Return Regulation: An Overview of the Discussion. *Bell Journal of Economics and Management Science* 1 (1):162-90.

Berndt, Ernst R. 1991. *The Practice of Econometrics: Classic and Contemporary*. Reading, MA: Addison-Wesley.

Berndt, Ernst R., and Eugene N. Savin. 1975. Estimation and Hypothesis Testing in Singular Equation Systems with Autoregressive Disturbances. *Econometrica* 43 (5-6, September-November):937-57.

Braeutigam, Ronald R., and John C. Panzar. 1993. Effects of the Change from Rate-of-Return to Price-Cap Regulation. *American Economic Review* 83 (2, May):191-89.

Christensen, Laurits R., Dale W. Jorgenson, and Lawrence J. Lau. 1971. Conjugate Duality and the Transcendental Logarithmic Function. *Econometrica* 39 (4, July):255-56.

Cowing, Tom. 1982. Duality and the Estimation of a Restricted Technology. In *Advances in Applied Micro-Economics*, ed. V. Kerry Smith. Greenwich, CT: JAI Press.

D'Aspremont, Claude, and Alexis Jacquemin. 1988. Cooperative and Noncooperative R&D in Duopoly with Spillovers. *American Economic Review* 78 (5, December):1133-37.

Davis, E. G. 1973. A Dynamic Model of the Regulated Firm with a Price Adjustment Mechanism. *Bell Journal of Economics and Management Science* 4 (1):270-82.

Dayan, David. 1975. Behavior of the Firm Under Regulatory Constraint: A Reexamination. *Industrial Organization Review* 3:31-76.

De Bondt, Raymond, Patrick Slaets, and Bruno Cassiman. 1992. The Degree of Spillovers and the Number of Rivals for Maximum Effective R&D. *International Journal of Industrial Organization* 10:35-54.

Eckert, Ross D. 1981. The Life Cycle of Regulatory Commissions. *Journal of Law and Economics* 24 (April):113-20.

Ellig, Jerry, and Michael Giberson. 1993. Scale, Scope, and Regulation in the Texas Gas Transmission Industry. *Journal of Regulatory Economics* 5:79-90.

FCC. 1992. *Federal Communications Commission Record* 7 (17, August 10 - August 21):5322-37.

Federal Power Commission v. Hope Natural Gas Co. 1944. *U.S.* 320:591.

Gayle, Gibson Jr. 1976. Statement of the Issue. *Baylor Law Review* 28 (4, Fall):771.

Gilligan, Thomas W., William J. Marshall, and Barry R. Weingast. 1989. Regulation and the Theory of Legislative Choice: The Interstate Commerce Act of 1887. *Journal of Law and Economics* 32 (April):35-61.

Granderson, Gerald. 1999. The Impact of Regulation on Technical Change. *Southern Economic Journal* 65 (4):807-22.

Hahn, Robert W., and John A. Hird. 1991. The Costs and Benefits of Regulation: Review and Synthesis. *Yale Journal on Regulation* 8:233-78.

Hilton, George W. 1972. The Basic Behavior of Regulatory Commissions. *American Economic Review* 62 (2, May):1972.

Hopper, Jack. 1976. A Legislative History of the Texas Public Utility Regulatory Act of 1975. *Baylor Law Review* 28 (4, Fall):777-822.

Hopper, Jack, and Eric Hartman. 1979. An A-minus from Wall Street, a D-minus from Consumers. *The Texas Observer*, July 27.

House Research Organization. 1996. *Power Struggle: Deregulating the Electricity Industry.* December 5. Texas House of Representatives.

Jewell, R. Todd, Daniel M. O'Brien, and Barry J. Seldon. Forthcoming. Media Substitution and Economies of Scale in Advertising. *International Journal of Industrial Organization.*

Kaestner, Robert, and Brenda Kahn. 1990. The Effects of Regulation and Competition on the Price of AT&T Intrastate Telephone Service. *Journal of Regulatory Economics* 2 (4, December):363-77.

Kahn, Alfred E. 1988. *The Economics of Regulation: Principles and Institutions.* Cambridge: MIT Press.

Kamien, Morton I., and Nancy L. Schwartz. 1982. *Market Structure and Innovation.* New York: Cambridge University Press.

Kaserman, David L., and John W. Mayo. 1995. *Government and Business: The Economics of Antitrust and Regulation.* Fort Worth: The Dryden Press.

Katz, Michael L. 1986. An Analysis of Cooperative Research and Development. *RAND Journal of Economics* 17 (4, Winter):527-43.

Klevorick, Alvin K. 1971. The "Optimal" Fair Rate of Return. *Bell Journal of Economics and Management Science* 2 (1):122-53.

Klevorick, Alvin K. 1973. The Behavior of a Firm Subject to Stochastic Regulatory Review. *Bell Journal of Economics and Management Science* 4 (1):57-88.

Kolbe, A. L., J. A. Read, and G. R. Hall. 1986. *The Cost of Capital: Estimating the Rate of Return for public Utilities*. Cambridge: The MIT Press.

Kumbhakar, Subal C. 1996. A Parametric Approach to Efficiency Measurement Using a Flexible Profit Function. *Southern Economic Journal* 63 (2, October):473-87.

Kwun, Younghan. 1987. Productivity and Regulation of electric Utilities: An Empirical Study. Masters Thesis. Austin, Texas: University of Texas at Austin.

Lucas, Robert E., Jr. 1988. On the Mechanics of Economic Development. *Journal of Monetary Economics* 22 (3, July):3-42.

Macauley, Molly. 1986. Out of Space? Regulation and Technical Change in Communication Satellites. *American Economic Review* 76 (2, May):280-84.

Mathios, Alan D., and Robert P. Rogers. 1989. The Impact of Alternative Forms of State Regulation of AT&T on Direct-Dial, Long-Distance Telephone Rates. *RAND Journal of Economics* 20 (3, Autumn):437-52.

Munn v. Illinois. 1887. *U.S.* 94:113.

National Association of Regulatory Utility Commissioners. 1996. *Profiles of Regulatory Agencies of the United States and Canada*. Yearbook 1995-1996. Washington D.C.

Nebbia v. New York. 1934. *U.S.* 291:502.

Nelson, Randy A. 1984. Regulation, Capital Vintage, and Technical Change in the Electric Utility Industry. *Review of Economic and Statistics* 66 (1, February):59-69.

Nelson, Richard R., and Sidney G. Winter. 1982. *An Evolutionary Theory of Economic Change*. Cambridge: Harvard University Press.

Okuguchi, Koji. 1975. The Implications of Regulation for Induced Technical Change: Comment. *Bell Journal of Economics* 6 (2, Autumn):703-05.

Peltzman, Sam. 1976. Toward a More General Theory of Regulation. *Journal of Law and Economics* 19.

Phillips, Charles F., Jr. 1988. *The Regulation of Public Utilities: Theory and Practice*. Second. Arlington, Virginia: Public Utilities Reports.

Phillips, Charles F., Jr. 1993. *The Regulation of Public Utilities: Theory and Practice*. Third. Arlington, Virginia: Public Utilities Reports.

Pleitz, Dan, and Robert Randolph Little. 1976. Municipalities and the Public Utility Regulatory Act. *Baylor Law Review* 28 (4, Fall):977-98.

Posner, Richard A. 1974. Theories of Economic Regulation. *Bell Journal of Economics and Management Science* 5 (2, Autumn):335-58.

Posner, Richard A. 1975. The Social Costs of Monopoly and Regulation. *Journal of Political Economy* 83 (4, August):807-27.

Poyago-Theotoky, Joanna. 1995. Equilibrium and Optimal Size of a Research Joint Venture in an Oligopoly with Spillovers. *The Journal of Industrial Economics* 63 (2, June):209-26.

Public Utility Commission of Texas. 1999a. *Scope of Competition in Telecommunications Market in Texas*. January. Report to the 76th Texas Legislature. Public Utility Commission of Texas.

Public Utility Commission of Texas. 1999b. *The Scope of Competition in the Electric Industry in Texas*. January. Report to the 76th Texas Legislature. Public Utility Commission of Texas.

Romer, Paul M. 1986. Increasing Returns and Long-Run Growth. *Journal of Political Economy* 94 (5, October):1002-37.

Romer, Paul M. 1994. The Origins of Endogenous Growth. *Journal of Economic Perspectives* 8 (1, Winter):3-22.

Scherer, F. M. 1992. Schumpeter and Plausible Capitalism. *Journal of Economic Literature* 30 (3, September):1416-33.

Schumpeter, Joseph A. 1939. *Business Cycles: A Theoretical, Historical and Statistical Analysis of the Capitalist Process*. New York: McGraw-Hill.

Schumpeter, Joseph A. 1950. *Capitalism, Socialism and Democracy*. New York: Harper Torchbooks.

Sherman, Roger. 1983. Is Public-Utility Regulation Beyond Hope? In *Current Issues in Public-Utility Economics: Essays in Honor of James C. Bonbright*, ed. Albert Danielsen and David R. Kamerschen. Lexington, Massachusetts: Lexington Books.

Sherman, Roger. 1985. The Averch and Johnson Analysis of Public Utility Regulation Twenty Years Later. *Review of Industrial Organization* 2 (2):178-93.

Smith, V. Kerry. 1974. The Implications of Regulation for Induced Technical Change. *Bell Journal of Economics and Management Science* 5 (2, Autumn):623-36.

Smith, V. Kerry. 1975. The Implications of Regulation for Induced Technical Change: Reply. *Bell Journal of Economics* 6 (2, Autumn):706-07.

Smyth v. Ames. 1898. *U.S.* 169:466.

Southwestern Bell Telephone Co. v. Missouri Public Service Commission. 1923. *U.S.* 262:603.

Stigler, George J. 1971. The Theory of Economic Regulation. *Bell Journal of Economics and Management Science* 2 (1, Spring).

Suzumura, Kotato. 1992. Cooperative and Noncooperative R&D in an Oligopoly with Spillovers. *American Economic Review* 82 (5, December):1307-20.

Takayama, Akira. 1969. Behavior of the Firm Under Regulatory Constraint. *American Economic Review* 59 (3):255-60.

Thompson, Howard E. 1991. *Regulatory Finance: Financial Foundations of Rate of Return Regulation*. Boston: Kluwer Academic Publishers.

Train, Kenneth E. 1991. *Optimal Regulation: The Economic Theory of Natural Monopoly*. Cambridge, Massachusetts: MIT Press.

Viscusi, W. Kip, John M. Vernon, and Joseph E. Harrington, Jr. 1997. *Economics of Regulation and Antitrust*. Second. Cambridge: MIT Press.

Wabash, St. Louis and Pacific Railway Company v. Illinois. 1886. *U.S.* 118:557.

Wellisz, Stanislaw H. 1963. Regulation of Natural Gas Pipeline Companies: An Economic Analysis. *Journal of Political Economy* 71 (February):30-43.

Westfield, Fred M. 1965. Regulation and Conspiracy. *American Economic Review* 55 (3, June):424-43.

Yi, Sang-Seung. 1996. The Welfare Effects of Cooperative R&D in Oligopoly with Spillovers. *Review of Industrial Organization* 11:681-89.

Zajac, E. E. 1970. A Geometric Treatment of Averch-Johnson's Behavior of the Firm Model. *American Economic Review* 60 (1):117-25.

Zellner, Arnold. 1962. An Efficient Method for Estimation Seemingly Unrelated Regressions and Tests for Aggregation Bias. *Journal of the American Statistical Association* 57 (June):5481-68.

Zellner, Arnold. 1963. Estimators for Seemingly Unrelated Regression Equations: Some Exact Finite Sample Estimates. *Journal of the American Statistical Association* 58 (December):977-92.

Index

Printed and bound by CPI Group (UK) Ltd, Croydon, CR0 4YY

21/10/2024

01777082-0013